Proceedings
of the Steklov Institute of Mathematics

Editors

T. P. Žukova G. A. Smirnova
R. G. Geršinskaja A. I. Kac

Number 115 (1971)

Mathematical Questions in the Theory of Wave Diffraction. I

edited by

V. M. Babič

AMERICAN MATHEMATICAL SOCIETY
Providence, Rhode Island
1974

Академия Наук
Союза Советских Социалистических Республик

ТРУДЫ
ордена Ленина
МАТЕМАТИЧЕСКОГО ИНСТИТУТА
имени В. А. СТЕКЛОВА
CXV

Ответственный редактор
В. М. БАБИЧ

МАТЕМАТИЧЕСКИЕ ВОПРОСЫ ТЕОРИИ ДИФРАКЦИИ И РАСПРОСТРАНЕНИЯ ВОЛН. 1

Издательство „Наука"
Ленинградское Отделение
Ленинград 1972

AMS (*MOS*) *subject classifications* (1970).
Primary 34B25, 35B40, 35J05, 35L05, 35L20, 35R30, 45K05, 73D25, 73D99, 78A40, 78A45; Secondary 30A88, 34E99, 35B40, 44A15, 44A35, 44A50, 45D05, 73D99, 73N99, 76Q05, 78A45, 86A15.

Library of Congress Cataloging in Publication Data CIP

Main entry under title:

Mathematical questions in the theory of wave diffraction.

(Proceedings of the Steklov Institute of Mathematics, no. 115)
Translation of Matematicheskie voprosy teorii difraktsii i rasprostraneniia voln.
1. Wave-motion, Theory of. 2. Differential equations, Partial. I. Babich, V. M., ed. II. Series: Akademiia nauk SSSR. Matematicheskii institut. Proceedings, no. 115.
QA1.A413 no. 115 510'.8s [531'.1133] 74-2362
ISBN 0-8218-3015-5

PROCEEDINGS OF THE STEKLOV INSTITUTE OF MATHEMATICS
IN THE ACADEMY OF SCIENCES OF THE USSR

(Труды математического института им. В. А. Стеклова, т. CXV, 1971)

TABLE OF CONTENTS

RAYLEIGH WAVES IN A SURFACE WAVEGUIDE

A. G. ALENICYN

UDC 517 944/947

Abstract. This paper considers stationary high-frequency Rayleigh waves in a halfspace $z \geqslant 0$ that is filled with an elastic medium whose parameters depend on z. The velocity $v_s(z)$ of the transverse wave and the velocity $v_p(z)$ of the longitudinal wave are monotonically increasing up to a certain value of z, after which they are constant. We study the region of phase velocities from zero to max $v_s(z)$. It is shown that with increasing frequency an "accumulation" of the plots of the phase velocities of the individual waves occurs in the vicinity of $v_s(0)$ and $v_p(0)$. For this investigation we construct a uniform asymptotic expression for the solutions of a system of ordinary differential equations in an interval containing two turning points.

Figures: 3. Bibliography: 5 items.

§1. Asymptotic formulas for the auxiliary system

We shall consider the problem of natural oscillations of an elastic isotropic halfspace $z \geqslant 0$ with parameters $\lambda(z)$, $\mu(z)$ and $\rho(z)$ which have three continuous derivatives for $z \geqslant 0$. In this paper we shall use the same notation as in [1].

In contrast to [1], here the parameters $\lambda(z)$, $\mu(z)$ and $\rho(z)$ are such that the velocities $v_p(z)$ of the longitudinal wave and $v_s(z)$ of the transverse wave are monotonically increasing in the interval $z \in [0, A]$ for

$$v_l'(z) > 0, \quad 0 \leqslant z < A, \quad l = p, s, \tag{1.1}$$

whereas for $z \geqslant A$ the parameters λ, μ and ρ (and hence also $v_l(z)$) are constant (Figure 1).

We shall also assume that

$$v_s(A) > v_p(0). \tag{1.2}$$

We shall study the region $0 < v < v_s(A)$, i.e. the discrete part of the spectrum of the Sturm-Liouville problem [1] (formulas (1.5)–(1.7)). Under the condition (1.2) the discrete spectrum is located in an interval containing $v_p^2(0)$.

As in [1], we seek an asymptotic expression, when $k \to \infty$, of the solutions of the auxiliary system

$$\mathbf{Z}' = (kH + K)\mathbf{Z}. \tag{1.3}$$

This asymptotic expression can be constructed with the aid[1] of the reference

AMS 1970 *subject classifications*. Primary 73D99, 34B25; Secondary 34E99.

1) Here we can avoid using the method of splitting [4], which requires infinitely smooth coefficients of the system.

method in exactly the same way as in [2] and [3]. In fact, the set of eigenvalues of the matrix $H(z,\sigma)$ decomposes into two pairs of numbers $[\pm m_p(z,\sigma), \pm m_s(z,\sigma)]$ such that $|m_p| \neq |m_s|$ if $\sigma > 0$. Hence there exists [4] a smooth nonsingular transformation $\mathbf{Z} = T(z,\sigma)\mathbf{X}$ that diagonalizes H:

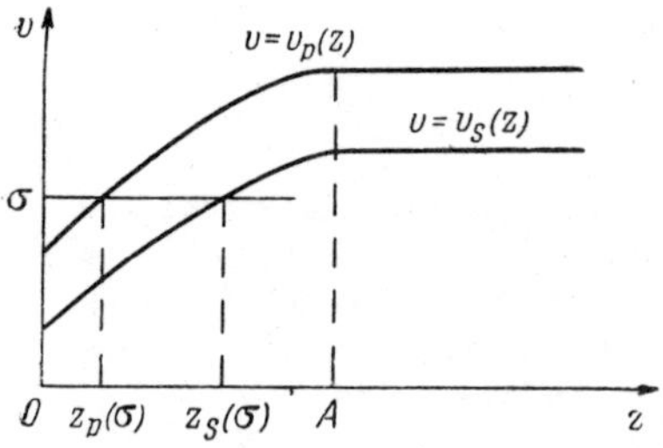

Figure 1

$$T^{-1}HT = [\tilde{H}_p,\ \tilde{H}_s], \tag{1.4}$$

$$\tilde{H}_p = \begin{pmatrix} 0 & 1 \\ m_p^2 & 0 \end{pmatrix}, \quad \tilde{H}_s = \begin{pmatrix} 0 & 1 \\ m_s^2 & 0 \end{pmatrix}. \tag{1.5}$$

The matrix $T(z,\sigma)$ has the form

$$T(z,\sigma) = \begin{pmatrix} -1 & 0 & 0 & 1 \\ 0 & 1 & -1 & 0 \\ 0 & -1 & m_s^2 & 0 \\ m_p^2 & 0 & 0 & -1 \end{pmatrix} \tag{1.6}$$

The reduced system is

$$\begin{gathered} \mathbf{X}' = (k\tilde{H} + \tilde{K})\mathbf{X}, \\ \tilde{K} = T^{-1}KT - T^{-1}T'. \end{gathered} \tag{1.7}$$

Let us consider an auxiliary quasidiagonal matrix

$$W(z, k, \sigma) = [W_p,\ W_s], \tag{1.8}$$

$$W_l = \begin{pmatrix} w_{1l} & w_{2l} \\ \mathring{w}_{1l} & \mathring{w}_{2l} \end{pmatrix}, \quad \mathring{f}k = \frac{d}{dz} f, \quad l = p,\ s,$$

$$w_{1l} = \frac{2k^{1/6}}{\sqrt{\varphi_l'}} v\,(k^{2/3}\varphi_l), \quad w_{2l} = \frac{k^{1/6}}{\sqrt{\varphi_l'}} u\,(k^{2/3}\varphi_l),$$

$$\varphi_l = \varphi_l(z,\sigma) = \left(\frac{3}{2}\int_{z_l(\sigma)}^{z} |m_l(x,\sigma)|\,dx\right)^{2/3} \operatorname{sign}[z - z_l(\sigma)].$$

Here $u(t)$ and $v(t)$ are Airy-Fok functions, and $z_p(\sigma)$ ($z_s(\sigma)$) is a longitudinal (transverse) turning point, i.e., $m_p(z_p,\sigma) = 0$ ($m_s(z_s,\sigma) = 0$). Under the conditions (1)–(2) there can be in the interval $0 \leqslant z \leqslant A$ not more than one turning point of each type; moreover, these will be simple turning points, since

$$\frac{d}{dz} m_l^2(z,\sigma) \neq 0. \tag{1.9}$$

It is evident that (1.8) is defined for all z only if $v_p(0) \leqslant \sigma \leqslant v_s(A)$. By continuing $v_p(z)$ and $v_s(z)$ to negative z with preservation of the smoothness class and of condition (1.1) and the natural condition $v_p(z) > v_s(z)$, it will be possible to consider (1.8) for all $0 < \sigma < v_s(A)$ and $z \geqslant 0$.

Since the w_{jl} satisfy the differential equation

$$-w''_{jl} + k^2 m_l^2 (z, \sigma) w_{jl} + r_l (z, \sigma) w_{jl} = 0 \qquad (l = p,\ s;\ j = 1,\ 2), \quad (1.10)$$

where $r_l(z,\sigma) = \psi''_l/\psi_l$, $\psi_l = [\varphi'_l(z,\sigma)]^{-1/2}$, it follows that (1.8) will be the fundamental matrix of the system

$$\mathbf{W}' = (k\tilde{H} + k^{-1}M)\mathbf{W},$$

$$M = [M_p,\ M_s],\quad M_l = \begin{pmatrix} 0 & 0 \\ r_l & 0 \end{pmatrix}. \quad (1.11)$$

The system (1.11) differs from (1.7) only by a term of the order of k^0, and (1.11) can be called a reference system for (1.7). For the proof of the desired asymptotic formulas it is necessary, however, that the systems coincide up to k^0th order inclusive; then the kernel of the corresponding Volterra-type integral equation [2, 3] will be of the order k^{-1}, as required. For this purpose we shall construct the following approximation for the desired vector **X**.

As in [2] and [3] we shall construct a formal series

$$X = \sum_{\nu=0}^{\infty} P_\nu (z, \sigma) k^{-\nu} W, \quad (1.12)$$

with the coefficients $P_\nu(z,\sigma)$ being infinitely smooth if $\lambda(z)$, $\mu(z)$ and $\rho(z)$ are such. Under the smoothness conditions adopted here, it is possible to construct only P_0, P_1 and P_2 in the form of continuous functions. Thus

$$P_0 = (\rho(0)/\rho(z))^{1/2} [1,\ 1,\ 1,\ 1]. \quad (1.13)$$

As in [3] (formula (3.38)), we have an asymptotic formula for $k \to \infty$:

$$Z(z, k, \sigma) = T(z, \sigma)[P_0(z, \sigma) + k^{-1}P_1(z, \sigma) + O(k^{-1})] W(z, k, \sigma)$$

$$= [T(z, \sigma) P_0(z, \sigma) + O(k^{-1})] W(z, k, \sigma)$$

$$= (\rho(0)/\rho(z))^{1/2} [T(z, \sigma) + O(k^{-1})] W(z, k, \sigma). \quad (1.14)$$

This formula is uniform for $0 \leqslant z \leqslant A$ and $0 < \varepsilon \leqslant \sigma \leqslant v_s(A) - \varepsilon$. It is easy to show that the constancy of λ, μ and ρ for $z \geqslant A$ is a sufficient condition of validity of (1.14) for $0 \leqslant z < +\infty$ (in fact, instead of the condition of constancy of λ, μ and ρ it is possible to impose conditions on their behavior for $z \to \infty$ such that the kernel of the corresponding integral equation will be integrable on a halfaxis; but this generalization will not be considered by us in view of its laboriousness).

§2. Dispersion curves

As usual, the dispersion curves denote the curves of the solutions $\sigma = \sigma(k)$ of the dispersion (characteristic) equation of the problem

$$\Delta(\sigma, k) = 0, \tag{2.1}$$

where, as in [1], we use the notation

$$\Delta(\sigma, k) = D_p E_s - E_p D_s, \tag{2.2}$$

$$D_l = k(Z_3^l - Z_2^l)\big|_{z=0}, \quad E_l = k\left(Z_4^l + \frac{\lambda}{\nu} Z_1^l\right)\Big|_{z=0}, \tag{2.3}$$

$$l = p,\ s; \quad \nu = \lambda + 2\mu.$$

Here $\mathbf{Z}^p$ and $\mathbf{Z}^s$ are two linearly independent solutions of (1.3) that decrease as $z \to \infty$, $0 < \sigma < v_s(A)$. It follows from (1.14) that $\mathbf{Z}^p$ is the first (and $\mathbf{Z}^s$ the third) column vector of the matrix Z.

Let us calculate D_l and E_l :

$$k^{-1}D_p = (-2 + O_1)\mathring{w}_{1p} + O_1 w_{1p}, \quad k^{-1}D_s = (1 + m_s^2 + O_1) w_{1s} + O_1 \mathring{w}_{1s}; \tag{2.4}$$

$$k^{-1}E_p = \mu(1 + m_s^2 + O_1) w_{1p} + O_1 \mathring{w}_{1p}, \quad k^{-1}E_s = (-2\mu + O_1)\mathring{w}_{1s} + O_1 w_{1s}. \tag{2.5}$$

In the formulas (2.4)–(2.5) and below, the symbol O_1 replaces $O(k^{-1})$. The estimates in (2.4)–(2.5) are uniform for $0 < \varepsilon \leqslant \sigma \leqslant v_s(A) - \varepsilon$; all the quantities are taken for $z = 0$.

Thus we have obtained a general representation for the dispersion equation that is valid for $0 < \varepsilon \leqslant \sigma \leqslant v_s(A) - \varepsilon$:

$$[(1 + m_s^2)^2 + O_1] w_{1p} w_{1s} - (4 + O_1)\mathring{w}_{1p}\mathring{w}_{1s} = k^{-1} O\{|w_{1p}\mathring{w}_{1s}| + |\mathring{w}_{1p} w_{1s}|\}. \tag{2.6}$$

Let us study separately the solutions of equation (2.6) in the following regions:

$$1)\ 0 < \sigma < v_{s0} \equiv v_s(0);\quad 2)\ \sigma \approx v_{s0};\quad 3)\ v_{s0} < \sigma < v_{p0} \equiv v_p(0);$$

$$4)\ \sigma \approx v_{p0};\quad 5)\ v_{p0} < \sigma < v_s(A).$$

1. $0 < \varepsilon \leqslant \sigma \leqslant v_{s0} - \varepsilon$. Here

$$w_{1l} = \{[m_l(0, \sigma)]^{-1/2} + O_1\} \exp\left(-k \int_{z_l(\sigma)}^{0} m_l(z, \sigma)\, dz\right), \tag{2.7}$$

$$\mathring{w}_{1l} = \{-[m_l(0, \sigma)]^{1/2} + O_1\} \exp\left(-k \int_{z_l(\sigma)}^{0} m_l(z, \sigma)\, dz\right). \tag{2.8}$$

Equation (2.6) goes over into the well-known equation for proper Rayleigh waves [5], similar to classical Rayleigh waves in an homogeneous halfspace:

$$(1+m_s^2)^2 - 4m_s m_p = O(k^{-1}). \tag{2.9}$$

Hence $\sigma = \sigma_R + O(k^{-1})$, where $\sigma_R > 0$ is a root of the equation (2.9) with a zero right-hand side.

2. $\sigma \approx v_{s0}$. The formulas (2.7)–(2.8) hold only for $l = p$. It is easy to see that

$$\begin{gathered} \varphi_s(0,\sigma) = -A_s(\sigma - v_{s0})[1 + O(\sigma - v_{s0})], \\ A_s = (2n_s)^{1/3}(v_s')^{-2/3}|_{z=0} > 0. \end{gathered} \tag{2.10}$$

Let $t \equiv -A_s(\sigma - v_{s0})k^{2/3} = O(1)$. Then

$$\begin{gathered} w_{1s} = \frac{2k^{1/6}}{\sqrt{\varphi_s'(0, v_{s0})}}[v(t) + O(k^{-2/3})], \\ \mathring{w}_{1s} = k^{-1/3}2k^{1/6}\sqrt{\varphi_s'(0, v_{s0})}\,[v'(t) + O(k^{-2/3})]. \end{gathered} \tag{2.11}$$

In the region $\sigma - v_{s0} = O(k^{-2/3})$ the equation (2.6) assumes the form

$$\begin{gathered} v(t) + B_s k^{-1/3} v'(t) + O(k^{-2/3}) = 0, \\ B_s = 4\sqrt{1-\gamma_0}\,\varphi_{s0}', \quad \varphi_{s0}' = -\sqrt[3]{n_s^2 (v_s^2)'}\,|_{z=0} < 0, \\ \gamma_0 = v_{s0}^2 n_{p0}^2 < \frac{1}{2}. \end{gathered} \tag{2.12}$$

Hence we obtain

$$\sigma_n = v_{s0} + \alpha_s \varkappa_n k^{-2/3} - \beta_s k^{-1} + O(k^{-4/3}), \tag{2.13}$$

where the $\kappa_n > 0$ are the zeros of the function $v(-k)$; $n = 1, 2, \dots, n_0 < \infty$; $\alpha_s = (0.5 v_{s0})^{1/3}(v_{s0}')^{2/3} > 0$, $v_{s0}' = v_s'(0)$ and $\beta_s = 4\sqrt{1-\gamma_0}\, v_{s0}' > 0$.

In this case it is possible to refer to solutions of transverse waveguide type.

3. $v_{s0} + \varepsilon \leqslant \sigma \leqslant v_{p0} - \varepsilon$.

$$w_{1s} = 2|m_s(0,\sigma)|^{-1/2}[\sin\Psi_s + O_1], \tag{2.14}$$

$$\mathring{w}_{1s} = -2|m_s(0,\sigma)|^{1/2}[\cos\Psi_s + O_1], \tag{2.15}$$

$$\Psi_s = \Psi_s(k,\sigma) = k\int_0^{z_s(\sigma)} |m_s(z,\sigma)|\,dz + \frac{\pi}{4}.$$

The equation (2.6) has the form

$$(1+m_s^2)^2 \sin\Psi_s - 4m_p|m_s|\cos\Psi_s = O_1, \tag{2.16}$$

or

$$\sin\Psi_s = f(\sigma)[\cos\Psi_s + O_1], \tag{2.17}$$

where $f(\sigma) = 4m_p|m_s|(1 + m_s^2)^{-2}$, $0 < f(\sigma) \leqslant +\infty$. Hence we obtain $k = k_l(\sigma)$ for $l \to \infty$:

$$k_l(\sigma) = [F_s(\sigma)]^{-1}\left[\left(l - \frac{1}{4}\right)\pi + \operatorname{arc\,tg} f(\sigma) + O(l^{-1})\right], \tag{2.18}$$

$$F_s(\sigma) = \int_0^{z_s(\sigma)} |m_s(z, \sigma)|\, dz > 0.$$

We can see that the distance $\Delta k_l(\sigma)$ between neighboring curves for fixed σ is equal in the first approximation to

$$\Delta k_l(\sigma) = \pi/F_s(\sigma). \tag{2.19}$$

4. $\sigma \approx v_{p0}$. For w_{1p} and $\overset{\circ}{w}_{1p}$ the formulas (2.11) remain valid if we effect the change of variables $s \to p$; for w_{1s} and $\overset{\circ}{w}_{1s}$ we can use (2.14)–(2.15). Equation (2.1) assumes the form

$$(\sin \Psi_s + O_1)\, v(t) + B_p k^{-1/3} (\cos \Psi_s + O_1)\, v'(t) = O(k^{-2/3}), \tag{2.20}$$

$$B_p = 4\sqrt{\gamma_0^{-1} - 1}\,(2 - \gamma_0^{-1})^{-2} \varphi'_{p0}, \quad \varphi'_{p0} = -\sqrt[3]{n_p^2 (v_p^2)'}\,\big|_{z=0} < 0,$$

$$-t = A_p(\sigma - v_{p0})\, k^{2/3} = O(1), \quad A_p = (2n_p)^{1/3} (v'_p)^{-2/3}\big|_{z=0} > 0.$$

This equation is of the same type as the equations studied in [1] and [2]. If $|v(t)| \geqslant \delta > 0$, then $\tan \Psi_s = O(k^{-1/3})$, i.e. roughly the same as in case 3. If $|\sin \Psi_s| \geqslant \delta > 0$, then $v(t) = O(k^{-1/3})$, which yields solutions of the form (2.13) corresponding to a "longitudinal waveguide." The term of the order $k^{-1/3}$ in (2.20) removes the intersection of curves, replacing it by quasi-intersection. In contrast to [1] and [2], the minimum distance between neighboring curves is not exponentially small, but of the order k^{-1}.

The author assumes that this is the first known case of nonexponential quasi-intersection.

5. $v_{p0} + \varepsilon \leqslant \sigma \leqslant v_s(A) - \varepsilon$. The dispersion equation reduces to the form

$$\sin \Psi_s \sin \Psi_p + O_1 = g(\sigma) \cos \Psi_s \cos \Psi_p, \tag{2.21}$$

$$\Psi_p = kF_p(\sigma) + \frac{\pi}{4}, \quad F_p(\sigma) = \int_0^{z_p(\sigma)} |m_p(z, \sigma)|\, dz > 0,$$

$$g(\sigma) = 4|m_s m_p|(1 + m_s^2)^{-2}, \quad 0 < \delta \leqslant g(\sigma) \leqslant c < \infty.$$

In connection with (2.21) it is natural to consider the first approximation equation

$$\sin \Psi_s \sin \Psi_p = g(\sigma) \cos \Psi_s \cos \Psi_p. \tag{2.22}$$

It is evident that (2.22) is equivalent to the equation

$$\sin kF_+(\sigma) = h(\sigma) \cos kF_-(\sigma), \tag{2.23}$$

where

$$F_{\pm}(\sigma) = F_s(\sigma) \pm F_p(\sigma) > 0,$$
$$h(\sigma) = \frac{g(\sigma) - 1}{g(\sigma) + 1}, \quad |h(\sigma)| < 1.$$

In the particular case $g = 1$ we obtain

$$\sin kF_+ = 0, \tag{2.24}$$
$$k_l(\sigma) = l\pi/[F_s(\sigma) + F_p(\sigma)], \tag{2.25}$$

and in the first approximation the distance between neighboring curves will be

$$\Delta k_l(\sigma) = \frac{\pi}{F_s(\sigma) + F_p(\sigma)}, \tag{2.26}$$

which is much less than according to formula (2.19)

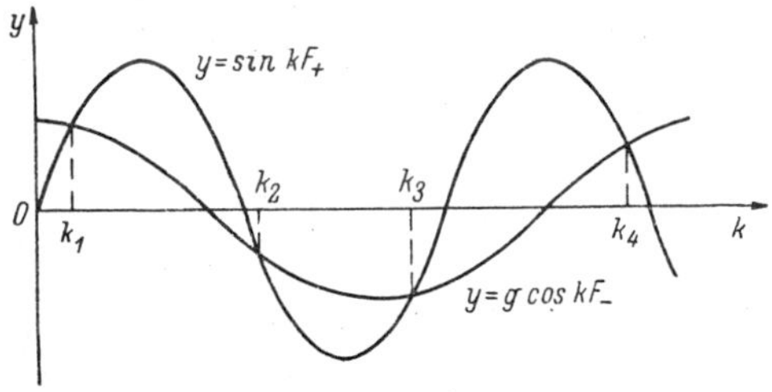

Figure 2

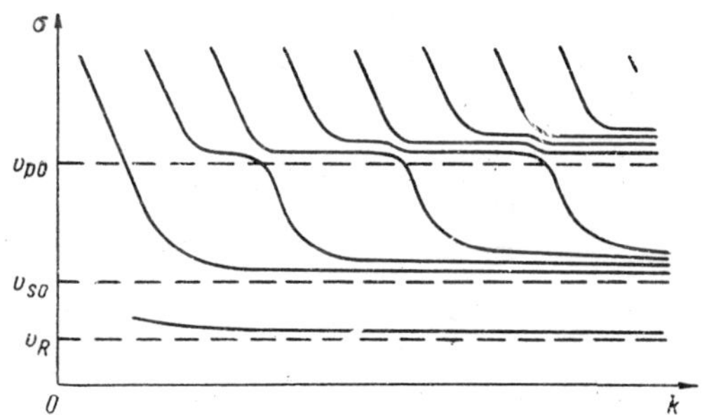

Figure 3

In the general case $g(\sigma) \neq 1$, the quantities $k_l(\sigma)$ are not calculated explicitly, but can be obtained graphically (Figure 2). It is evident that in any interval $k \in I = (a, b)$ the number of roots of the equation (2.23) can differ from the number of roots of the equation (2.24) by not more than 1. The average (in the interval I) distance between the roots of (2.23) approaches (2.25) when I increases. For $k \gg 1$, the equation (2.21) has a similar behavior.

In Figure 3 we plot schematically a family of dispersion curves for $k \gg 1$, $0 < \sigma < v_{s\max}$, and $g(\sigma) \doteq 1$. It can be seen that when $k \to \infty$, we have in the region $\sigma \approx v_{p0}$, as well as $\sigma \approx v_{s0}$, some sort of accumulation of dispersion curves. The physical nature of this effect is not very clear.

BIBLIOGRAPHY

1. A.G. Alenicyn, *Rayleigh waves in an inhomogeneous elastic half-space of waveguide type*, Prikl. Mat. Meh. **31** (1967), 222–229 = J. Appl. Math. Mech. **31** (1967), 244–251.
2. ———, *Rayleigh waves in a nonhomogeneous elastic slab*, Prikl. Mat. Meh. **28** (1964), 880–888 = J. Appl. Math. Mech. **28** (1964), 1067–1076. MR **32** #3361.
3. ———, *Asymptotic behavior of the solutions of linear systems of ordinary differential equations*, Problems of Diffraction and Wave Propagation, no. 7, Izdat. Leningrad. Univ., Leningrad, 1968, pp. 3–18. (Russian) RŽ Mat. 1968 164.
4. S.F. Feščenko, N.I. Škil′ and L.D. Nikolenko, *Asymptotic methods in the theory of linear differential equations*, "Naukova Dumka", Kiev, 1966. (Russian) MR **33** #6057.
5. A.G. Alenicyn, *Rayleigh waves in a nonhomogeneous elastic half-space*, Prikl. Mat. Meh. **27** (1963), 547–550 = J. Appl. Math. Mech. **27** (1963), 816–822. MR **28** #1825.

Translated by:
L. Ebner

A FORMAL METHOD OF CONSTRUCTION OF A SHORT-WAVE ASYMPTOTIC EXPRESSION FOR GREEN'S FUNCTION

V. M. BABIČ

UDC 517 944/947

Abstract. A formal method of construction is presented for an asymptotic expression for Green's function that requires fewer calculations as compared to the method based on the solution of a reference problem and the method of comparison with a ray solution.
Bibliography: 7 items.

In this paper we construct a short-wave asymptotic expression for the solution of the problem of a point source of oscillations located near a convex curve S, on the side of this curve where waves of whispering-gallery type are not generated [1]; more precisely, for the solution of the equation

$$
\begin{gathered}
(\Delta_M + k^2)\, G(M,\ M_0;\ k) = -\delta(M - M_0), \\
M = M(x,\ y),\ M_0 = M_0(x_0,\ y_0),\ \Delta_M = \frac{\partial^2}{\partial x^2} + \frac{\partial^2}{\partial y^2},\ k > 0,
\end{gathered} \tag{1}
$$

with the boundary condition

$$
G\,|_{M \in S} = 0 \tag{2}
$$

(here $\delta(M - M_0)$ is Dirac's delta function, concentrated at the point M_0).

The function G, usually called Green's function, will be sought in the form of a superposition of functions that admit an analytic continuation with respect to k in the region $\operatorname{Im} k > 0$; for large k these functions are exponentially decreasing with the distance from the contour S. In this sense the function G constructed below will satisfy the classical principle of limiting absorption.

The formula for G, obtained here, is well known in connection with reference [2], where formulas for G are used that are entirely similar to our formulas.

Our derivation of the asymptotic expression for G is very simple and can be used in a wide range of cases, which is precisely the reason for writing this article. Let us note that all our constructions are formal and cannot be justified even at the "physical" level of mathematical rigor (yet in spite of such an "unreliable" derivation, our formulas for G are confirmed from various points of view; they are in agreement with Keller's formulas [3] for Green's function in the shadow region, with the estimates obtained in [4] and [5], and with the explicit expressions for G that could be obtained if S is a circle, an ellipse, or a parabola).

AMS 1970 *subject classifications*. Primary 35B40, 35J05; Secondary 78A45.

We shall assume that M and M_0 are located near the curve S. The points on S will be specified by the arc length s, and the points M lying outside S will be specified by a pair of numbers (s, n), where n is the distance between M and S along the normal, and s is the coordinate of the projection (along this normal) of the point M onto S. We shall assume that in this region $n \geqslant 0$.

Now equation (1) and the boundary conditions will assume the form

$$\frac{\partial}{\partial n}\left[\left(1+\frac{n}{\rho}\right)\frac{\partial G}{\partial n}\right]+\frac{\partial}{\partial s}\left[\left(1+\frac{n}{\rho}\right)^{-1}\frac{\partial G}{\partial s}\right]=-\delta(s-s_0)\,\delta(n-n_0), \tag{3}$$

$$G=G(M,\ M_0,\ k),\quad M=M(s,\ n),\quad M_0=M_0(s_0,\ n_0), \tag{4}$$
$$G\,|_{n=0}=0.$$

(here $\rho = \rho(s) > 0$ is the radius of curvature of S). It is easy to see [1] that for small n the expression

$$u_p^+(s,\ n,\ k)$$
$$=\rho^{-1/6}(s)\exp\left[iks+i\xi_p\left(\frac{k}{2}\right)^{1/3}\int_0^s\frac{ds}{\rho^{2/3}}+\frac{i}{6k^{1/3}}\frac{\rho'(s)}{\rho(s)}\nu^2\right]w_1\left(\xi_p-\left(\frac{2}{\rho(s)}\right)^{1/3}\nu\right) \tag{5}$$

(here $\nu = k^{2/3}n$, w_1 is Airy's function in Fok's definition [6], and ξ_p, $p = 0, 1, \dots$, $\xi_p = |\xi_p| \exp i\pi/3 \neq 0$ are its roots) satisfies to within the principal terms in $k(k \to \infty)$ the equation (3) with a zero right-hand side. For example, for $n = O(k^{-2/3})$ the terms of the order of k^2, $k^{4/3}$ and k in the left-hand side of the equation are cancelled. The residual will have the order $O(k^{2/3})$. Formula (5) will evidently satisfy the boundary condition (4).

Assuming that the time dependence of the "wave" (5) is described by a factor e^{-ikt}, it is easy to see that the corresponding surface of equal phases (the wave front) has (to within principal terms) the form $t = s$, i.e. the wave u_p^+ propagates at a velocity 1 towards increasing s. By formally replacing i by $-i$ in (5), we obtain a wave

$$u_p^-(s,\ n;\ k)$$
$$=\rho^{-1/6}(s)\exp\left[-iks-i\xi_p\left(\frac{k}{2}\right)^{1/3}\int_0^s\frac{ds}{\rho^{2/3}(s)}+\frac{i}{6k^{1/3}}\frac{\rho'(s)}{\rho(s)}\nu^2\right]w_1\left(\xi_p-\left(\frac{2}{\rho(s)}\right)^{1/3}\nu\right), \tag{6}$$

that propagates towards decreasing s.

For $s > s_0$ (or $s < s_0$) it is natural to seek Green's function $G(s, n; s_0, n_0; k)$ in the form of a superposition of waves that "travel" towards increasing (or decreasing) s, i.e.

$$G = \begin{cases} \sum_{p=0}^{\infty} \mu_p^+(s_0, n_0; k) u_p^+(s, n; k), & s > s_0; \\ \sum_{p=0}^{\infty} \mu_p^-(s_0, n_0; k) u_p^-(s, n; k), & s < s_0. \end{cases} \tag{7}$$

For the "reciprocity principle" to hold, i.e. for the identity

$$G(M, M_0; k) \equiv G(M_0, M; k),$$
$$M = M(s, n), \quad M_0 = M_0(s_0, n_0), \tag{8}$$

we shall write

$$\mu_p^{\pm}(s_0, n_0; k) = A_p u_p^{\mp}(s_0, n_0; k), \tag{9}$$

where $A_p = A_p(k)$ does not depend on s_0 and n_0. Now the expression (7) for Green's function assumes the form[1]

$$G(M, M_0; k) = \frac{e^{ik|s-s_0|}}{\sqrt{\rho(s_0)\rho(s)}} \sum_{p=0}^{\infty} A_p \exp\left\{\frac{i\xi_p k^{1/3}}{2^{1/3}} \left| \int_{s_0}^{s} \frac{ds}{\rho^{2/3}(s)} \right|\right.$$
$$\left. + \frac{i}{6k^{1/3}} \left(\frac{\rho'(s)}{\rho(s)} \nu^2 + \frac{\rho'(s_0)}{\rho(s_0)} \nu_0^2\right)\right\} w_1\left(\xi_p - \left(\frac{2}{\rho(s)}\right)^{1/3} \nu\right) w_1\left(\xi_p - \left(\frac{2}{\rho(s_0)}\right)^{1/3} \nu_0\right). \tag{10}$$

Let us substitute (10) into (3). In the left-hand side we can obtain $\delta(s - s_0)$ by twice differentiating, with respect to s, the expression

$$\exp ik|s - s_0| + i\xi_p \left(\frac{k}{2}\right)^{1/3} \left| \int_{s_0}^{s} \frac{ds}{\rho^{2/3}(s)} \right|;$$

here

$$\frac{d^2}{ds^2} |s - s_0| = 2\delta(s - s_0)$$

and

$$\frac{d^2}{ds^2} \left| \int_{s_0}^{s} \frac{ds}{\rho^{2/3}(s)} \right| = 2\rho^{-2/3}(s_0)\delta(s - s_0) + \operatorname{sign}(s - s_0) \frac{d}{ds} \rho^{-2/3}(s).$$

Let us equate the expressions containing $\delta(s - s_0)$ in the left and right sides of the formula. To within principal terms we obtain

---

[1] By a similar reasoning, a formula similar to (10) was obtained in [7], Part II, §3°. In [7] the expansion coefficients were obtained on the basis of entirely different considerations.

$$\frac{2ik}{[\rho(s_0)]^{1/3}}\sum_{p=0}^{\infty}A_p w_1\left(\xi_p-\left(\frac{2}{\rho(s_0)}\right)^{1/3}\nu\right)w_1\left(\xi_p-\left(\frac{2}{\rho(s_0)}\right)^{1/3}\nu_0\right)=-\delta(n-n_0); \tag{11}$$

$$\nu=k^{2/3}n;\quad \nu_0=k^{2/3}n_0.$$

By setting

$$\sigma=\left(\frac{2}{\rho(s_0)}\right)^{1/3};\quad \nu=\left(\frac{2}{\rho(s_0)}\right)^{1/3}k^{2/3}n;\quad \sigma_0=\left(\frac{2}{\rho(s_0)}\right)^{1/3};\quad \nu_0=\left(\frac{2}{\rho(s_0)}\right)^{1/3}k^{2/3}n_0$$

and using the relation

$$\delta(\lambda(n-n_0))=\frac{1}{\lambda}\delta(n-n_0),$$

we can rewrite (11) as follows:

$$2i\left(\frac{k}{2}\right)^{1/3}\sum_{p=0}^{\infty}A_p w_1(\xi_p-\sigma)\,w_1(\xi_p-\sigma_0)=-\delta(\sigma-\sigma_0). \tag{12}$$

By multiplying this equation by $w_1(\xi_q-\sigma)$, integrating from zero to infinity, and using the easily verified formula

$$\int_0^{\infty}w_1(\xi_p-\sigma)\,w_1(\xi_q-\sigma)\,d\sigma=\begin{cases}0, & p\neq q,\\ -[w'(\xi_p)]^2, & p=q,\end{cases} \tag{13}$$

we obtain

$$A_p=\frac{1}{2i}\left(\frac{2}{k}\right)^{1/3}[w'(\xi_p)]^{-2}. \tag{14}$$

Let us substitute this expression into the series (10):

$$G(M,M_0;k)=\frac{1}{2i}\left(\frac{2}{k}\right)^{1/3}\frac{e^{ik|s-s_0|}}{\sqrt[6]{\rho(s)\rho(s_0)}}\sum_{p=0}^{\infty}\frac{1}{[w'(\xi_p)]^2}\exp\frac{i\xi_p k^{1/3}}{2^{1/3}}\left|\int_{s_0}^{s}\frac{ds}{\rho^{2/3}(s)}\right|$$

$$\times\, w_1\left(\xi_p-\left(\frac{2}{\rho(s)}\right)^{1/3}\nu\right)w_1\left(\xi_p-\left(\frac{2}{\rho(s_0)}\right)^{1/3}\nu_0\right), \tag{15}$$

$$\nu=k^{2/3}n;\quad \nu_0=k^{2/3}n_0.$$

(Since we are interested only in the principal terms, we dropped in the exponent the term of order $k^{-1/3}$.)

Using the residue theorem, we can express the series (15) in the form of Fok's classical integral:

$$G(M,\ M_0;\ k)=\frac{i}{8\pi}\left(\frac{2}{k}\right)^{1/3}\left(\frac{1}{\rho(s)\,\rho(s_0)}\right)^{1/6}\int\limits_C\left[w_2(\xi-Y_<)-\frac{w_2(\xi)}{w_1(\xi)}w_1(\xi-Y_<)\right]$$

$$\times\, w_1(\xi-Y_>)\exp\left[i\left(\frac{k}{2}\right)^{1/3}\xi\left|\int\limits_{s_0}^{s}\frac{ds}{\rho^{2/3}(s)}\right|\right]e^{ik|s-s_0|}. \tag{16}$$

Here w_1 and w_2 are Airy functions [7], C is a contour that encircles in the positive direction the ray $\arg\xi=\pi/3$ on which the poles of the integrand are located, and $Y_<$ ($Y_>$) is the smallest (largest) of the expressions

$$\left(\frac{2}{\rho(s_0)}\right)^{1/3}\nu_0,\quad \left(\frac{2}{\rho(s)}\right)^{1/3}\nu.$$

Bibliography

1. V.S. Buldyrev, *Asymptotics of eigenfunctions of the Helmholtz equation for plane convex regions*, Vestnik Leningrad. Univ. **20** (1965), no. 22, 38–51. (Russian) MR **32** #8692.
2. V.S. Buslaev, *Formulas for the short-wave asymptotic behavior in the diffraction problem by convex bodies*, Vestnik Leningrad. Univ. **17** (1962), no. 13, 5–21. (Russian) MR **25** #5693.
3. J.B. Keller, *Diffraction by a convex cylinder*, Trans. IRE **AP-4** (1965), 312–321. MR **20** #641.
4. V.M. Babič, *An estimate for the Green's function of the Helmholtz equation in the shadow region*, Vestnik Leningrad. Univ. **20** (1965), no. 7, 5–15. (Russian) MR **32** #5001.
5. V.M. Babič and I.V. Olimpiev, *Estimates of field in the shadow region in the case of diffraction of a cylindrical wave by a bounded convex cylinder*, Proc. Third All-Union Sympos. Wave Diffraction, "Nauka", Leningrad, 1964. (Russian)
6. V.A. Fok, *Tables of Airy functions*, Naučno-Issled. Inst., Inform. Otd. **108** (1946).
7. V.A. Fok and L.A. Vaĭnšteĭn, *Transverse diffusion in the case of diffraction of short waves by a convex cylinder of smoothly varying curvature*. I, II, Radiotehn. i Èlektron. **8** (1963), 363–376, 377–388 = Radio Engrg. Electron. Phys. **8** (1963), 317–330, 330–341.

Translated by:
L. Ebner

DIFFRACTION OF A CYLINDRICAL ELASTIC WAVE ON A SEMI-INFINITE SCREEN

B. P. BELINSKIĬ AND D. P. KOUZOV

UDC 517 944/947

Abstract. The authors consider the problem of diffraction of elastic waves from a semi-infinite cut on whose edges either the displacement components vanish (rigid screen) or the components of the stress tensor vanish (crack). Unlike Fridman, Maue, and Filippov, who studied only the case of a plane incident wave, the authors consider a point source. For a general point source they find a convenient representation which makes it possible to reduce the solution of the diffraction problem to the solution of integral equations. Solving these equations by the factorization method, the authors obtain an exact solution for the diffraction problem.

Bibliography: 7 items.

We consider the problem of the stationary oscillations of an elastic plane in the presence of a semi-infinite screen. The field source is an external force applied at some point of the medium. First we seek (§1) a special representation for the field of a point force in an unbounded space (i.e. in the absence of a screen). We construct an exact solution for the case of a rigid screen (§2), and also for a semi-infinite cut in the elastic medium (§3). It seems that such problems have been considered only for the case of a plane incident wave [1–5].

We use the following notation: ρ is the density of the medium; ω is the frequency of oscillation; $\mathbf{u} = (u, v, w) = \mathbf{u}(x, y)$ is the displacement vector; k and K are wave numbers; $\mathbf{f}$ is the linear force density; σ_{yy}, σ_{xy}, σ_{zy} are the components of the stress tensor; σ is the Poisson coefficient; and E is Young's modulus. The time factor $e^{-i\omega t}$ is omitted throughout.

Consider an elastic medium whose motion is described in terms of the displacement vector $\mathbf{u}$. Assume that the processes in the medium are subject to the Lamé equation and (with a suitably chosen coordinate system) depend on the two spatial variables x and y.

At the point (x_0, y_0) the medium is subject to some external force which is a harmonic function of time. On the halfline $x < 0$, $y = 0$ the mechanical properties of the medium are violated. We will consider two types of violations: the presence of a nondeforming body on this halfline (a rigid screen; in this case we have the boundary condition $\mathbf{u} = 0$ for $x < 0$, $y = 0$), and the presence of a cut in the medium (here we have $\sigma_{yy} = \sigma_{xy} = \sigma_{zy} = 0$).

It is the aim of this paper to construct exact solutions for these diffraction problems. We will use the method of expanding the solution in plane waves [3, 6],

---

AMS 1970 *subject classifications*. Primary 73D25; Secondary 30A88.

after which the problems reduce to nonhomogeneous boundary value problems for analytic functions.

First (§1) we will find a suitable representation for the field of the incident wave.

§1. Field of a point force in the absence of a screen

Consider harmonic motion of a uniform unbounded elastic medium. The field source is an external force of constant linear density **f**, which is concentrated along the line $x = 0$, $y = 0$. The field components do not depend on the third coordinate z (plane problem). Over the entire plane (x,y) we assume that the Lamé equation holds:

$$\rho\omega^2\mathbf{u}+\frac{E(1-\sigma)}{(1-2\sigma)(1+\sigma)}\nabla\operatorname{div}\mathbf{u}-\frac{E}{2(1+\sigma)}\operatorname{rot}\operatorname{rot}\mathbf{u}$$
$$+\mathbf{f}\cdot\delta(x-x_0,\ y-y_0)=0. \tag{1.1}$$

Dividing term by term by $\rho\omega^2$, we can transform (1.1) to the following:

$$\mathbf{u}+\frac{1}{k^2}\nabla\operatorname{div}\mathbf{u}-\frac{1}{K^2}\operatorname{rot}\operatorname{rot}\mathbf{u}+\mathbf{F}\cdot\delta(x-x_0,\ y-y_0)=0. \tag{1.2}$$

Here

$$k^2=\frac{\rho\omega^2(1+\sigma)(1-2\sigma)}{E(1-\sigma)},\quad K^2=\frac{2\rho\omega^2(1+\sigma)}{E},\quad \mathbf{F}=\frac{\mathbf{f}}{\rho\omega^2}.$$

Here and below the wave numbers k and K are assumed to be complex; thus $\mathrm{Re}(k,K)>0$, $\mathrm{Im}(k,K)>0$. We seek the solution in the class of functions which decrease exponentially as $\sqrt{(x^2+y^2)}\to\infty$. For real k and K the solution is understood as the limit of the solution for complex k and K as $\mathrm{Im}(k,K)\to+0$ (principle of limiting absorption).

We write (1.2) in terms of its components:

$$u+\frac{1}{k^2}(u_{xx}+v_{xy})+\frac{1}{K^2}(u_{yy}-v_{xy})+F_x\cdot\delta(x-x_0,\ y-y_0)=0,$$
$$v+\frac{1}{k^2}(u_{xy}+v_{yy})+\frac{1}{K^2}(v_{xx}-u_{xy})+F_y\cdot\delta(x-x_0,\ y-y_0)=0, \tag{1.3}$$
$$w+\frac{1}{K^2}(w_{xx}+w_{yy})+F_z\cdot\delta(x-x_0,\ y-y_0)=0.$$

The solution of our problem may be obtained as a linear combination

$$\mathbf{u}=F_x\cdot\mathbf{u}_A+F_y\cdot\mathbf{u}_B+F_z\cdot\mathbf{u}_C \tag{1.4}$$

of the solutions of the following three problems: problem A: $\mathbf{F}=(1,0,0)$; problem B: $\mathbf{F}=(0,1,0)$; problem C: $\mathbf{F}=(0,0,1)$. It is clear that here $w_A=w_B=u_C=v_C=0$.

The solution of problem C is known, and problems A and B are solved in a similar manner. Here we will consider problem A; for brevity the subscript A is omitted in (1.9)–(1.19). For problem A the first two equations of (1.3) may be rewritten as follows:

$$Pu + Qv = -\delta(x - x_0,\ y - y_0), \quad Qu + Rv = 0, \tag{1.5}$$

where we have used the notation

$$\begin{aligned} P &= 1 + \frac{1}{k^2}\frac{\partial^2}{\partial x^2} + \frac{1}{K^2}\frac{\partial^2}{\partial y^2}, \\ Q &= \left(\frac{1}{k^2} - \frac{1}{K^2}\right)\frac{\partial^2}{\partial x \partial y}, \\ R &= 1 + \frac{1}{k^2}\frac{\partial^2}{\partial y^2} + \frac{1}{K^2}\frac{\partial^2}{\partial x^2}. \end{aligned} \tag{1.6}$$

The solution (u, v) is sought in the form

$$u = u_1 + u_3, \quad v = v_2 + v_4, \tag{1.7}$$

where

$$\begin{aligned} \mathbf{u}_1(x,\ y) &= \int_{-\infty}^{+\infty} \mathbf{p}_1(\lambda)\, e^{i\lambda(x - x_0) - \sqrt{\lambda^2 - k^2}\,|y - y_0|}\, d\lambda, \\ \mathbf{u}_2(x,\ y) &= \int_{-\infty}^{+\infty} \mathbf{p}_2(\lambda)\, e^{i\lambda(x - x_0) - \sqrt{\lambda^2 - k^2}\,|y - y_0|} \cdot \operatorname{sign}(y - y_0) \cdot d\lambda, \\ \mathbf{u}_3(x,\ y) &= \int_{-\infty}^{+\infty} \mathbf{p}_3(\lambda)\, e^{i\lambda(x - x_0) - \sqrt{\lambda^2 - K^2}\,|y - y_0|}\, d\lambda, \\ u_4(x,\ y) &= \int_{-\infty}^{+\infty} \mathbf{p}_4(\lambda)\, e^{i\lambda(x - x_0) - \sqrt{\lambda^2 - K^2}\,|y - y_0|} \cdot \operatorname{sign}(y - y_0)\, d\lambda, \\ \mathbf{p}_i &= (p_i,\ q_i) \qquad (t = 1,\ 2,\ 3,\ 4). \end{aligned} \tag{1.8}$$

The branches of the radicals are fixed by the requirements

$$\begin{aligned} &\lim_{\lambda \to \pm\infty} \operatorname{Im} \sqrt{\lambda^2 - k^2} = 0\, (\operatorname{Im} \lambda = 0), \\ &\lim_{\lambda \to \pm\infty} \operatorname{Im} \sqrt{\lambda^2 - K^2} = 0\, (\operatorname{Im} \lambda = 0). \end{aligned} \tag{1.9}$$

The cuts on the λ plane go from the points $\lambda = \pm k, \pm K$ to infinity without intersecting the real axis.

The choice of representation (1.7) for the desired solution (u, v) is related to the obvious fact that u is even and v is odd relative to the variable $y - y_0$. The functions p_1, p_3, q_2 and q_4 are unknown and are sought on the basis of (1.5). With the branches of the radicals thus chosen, the solution **u** automatically has the requisite behavior as $\sqrt{(x^2 + y^2)} \to \infty$.

It is easy to see that

$$Pu_1 + Qv_2 = \frac{2i}{k^2}\delta(y - y_0)\int_{-\infty}^{+\infty}\left\{-\sqrt{\lambda^2 - k^2}\,\frac{p_1(\lambda)}{K^2} + i\left(\frac{1}{k^2} - \frac{1}{K^2}\right)\lambda q_2(\lambda)\right\}$$

$$\times e^{i\lambda(x-x_0)}d\lambda + \int_{-\infty}^{+\infty}\left\{-p_1(\lambda)\cdot(\lambda^2 - k^2)\cdot\left(\frac{1}{k^2} - \frac{1}{K^2}\right) - q_2(\lambda)\cdot i\lambda\sqrt{\lambda^2 - k^2}\right.$$

$$\left.\times\left(\frac{1}{k^2} - \frac{1}{K^2}\right)\right\} e^{i\lambda(x-x_0) - \sqrt{\lambda^2 - k^2}\,|y - y_0|}\,d\lambda. \qquad (1.10)$$

If we set

$$p_1(\lambda) = -i\lambda\varphi_1(\lambda), \quad q_2(\lambda) = \sqrt{\lambda^2 - k^2}\,\varphi_1(\lambda), \qquad (1.11)$$

where $\varphi_1(\lambda)$ is a new unknown function, the first integral in (1.10) vanishes, and this expression becomes

$$Pu_1 + Qv_2 = \frac{2i}{k^2}\delta(y - y_0)\cdot\int_{-\infty}^{+\infty}\lambda\sqrt{\lambda^2 - k^2}\,\varphi_1(\lambda)\,e^{i\lambda(x-x_0)}\,d\lambda. \qquad (1.12)$$

It may similarly be shown that for condition (1.11)

$$Qu_1 + Rv_2 = \frac{2}{k^2}\delta'(y - y_0)\cdot\int_{-\infty}^{+\infty}\sqrt{\lambda^2 - k^2}\,\varphi_1(\lambda)\,e^{i\lambda(x-x_0)}\,d\lambda; \qquad (1.13)$$

and $Pu_3 + Qv_4$ is transformed similarly. Here we introduce the new unknown function $\varphi_2(\lambda)$, for which

$$p_3(\lambda) = \sqrt{\lambda^2 - K^2}\,\varphi_2(\lambda), \quad q_4(\lambda) = i\lambda\varphi_2(\lambda). \qquad (1.14)$$

It is easy to see that

$$Pu_3 + Qv_4 = -\frac{2}{k^2}\delta(y - y_0)\cdot\int_{-\infty}^{+\infty}(\lambda^2 - k^2)\,\varphi_2(\lambda)\,e^{i\lambda(x-x_0)}d\lambda, \qquad (1.15)$$

$$Qu_3 + Rv_4 = \frac{2}{k^2}\delta'(y - y_0)\cdot\int_{-\infty}^{+\infty}i\lambda\varphi_2(\lambda)\,e^{i\lambda(x-x_0)}d\lambda. \qquad (1.16)$$

Hence equations (1.5) become

$$\frac{2}{k^2}\cdot\delta(y-y_0)\cdot\int\limits_{-\infty}^{+\infty}\{i\lambda\sqrt{\lambda^2-k^2}\,\varphi_1(\lambda)-(\lambda^2-k^2)\,\varphi_2(\lambda)\}\,e^{i\lambda(x-x_0)}\,d\lambda$$

$$=-\frac{1}{2\pi}\,\delta(y-y_0)\cdot\int\limits_{-\infty}^{+\infty}e^{i\lambda(x-x_0)}\,d\lambda, \qquad (1.17)$$

$$\frac{2}{k^2}\cdot\delta'(y-y_0)\cdot\int\limits_{-\infty}^{+\infty}\{\sqrt{\lambda^2-k^2}\,\varphi_1(\lambda)+i\lambda\varphi_2(\lambda)\}\,e^{i\lambda(x-x_0)}\,d\lambda=0.$$

These equations are satisfied identically if we set

$$\frac{2i}{k^2}\lambda\sqrt{\lambda^2-k^2}\,\varphi_1(\lambda)-\frac{2}{k^2}(\lambda^2-k^2)\,\varphi_2(\lambda)=-\frac{1}{2\pi},$$

$$\frac{2}{K^2}\sqrt{\lambda^2-k^2}\,\varphi_1(\lambda)+\frac{2}{K^2}i\lambda\varphi_2(\lambda)=0, \qquad (1.18)$$

whence

$$\varphi_1(\lambda)=\frac{1}{4\pi}\cdot\frac{i\lambda}{\sqrt{\lambda^2-k^2}},\quad \varphi_2(\lambda)=-\frac{1}{4\pi}. \qquad (1.19)$$

As a result we obtain the following solution for problem A:

$$u_A(x,y)=\frac{1}{4\pi}\int\limits_{-\infty}^{+\infty}e^{i\lambda(x-x_0)}\left\{\frac{\lambda^2}{\sqrt{\lambda^2-k^2}}\,e^{-\sqrt{\lambda^2-k^2}\,|y-y_0|}-\sqrt{\lambda^2-K^2}\,e^{-\sqrt{\lambda^2-K^2}\,|y-y_0|}\right\}d\lambda,$$

$$v_A(x,y)=\frac{i}{4\pi}\int\limits_{-\infty}^{+\infty}e^{i\lambda(x-x_0)}\left\{\lambda\left(e^{-\sqrt{\lambda^2-k^2}\,|y-y_0|}-e^{-\sqrt{\lambda^2-K^2}\,|y-y_0|}\right)\right\}\operatorname{sign}(y-y_0)\,d\lambda,$$

$$w_A(x,y)=0. \qquad (1.20)$$

The solution of problem B is similar:

$$u_B(x,y)=\frac{i}{4\pi}\int\limits_{-\infty}^{+\infty}e^{i\lambda(x-x_0)}\left\{\lambda\left(e^{-\sqrt{\lambda^2-k^2}\,|y-y_0|}-e^{-\sqrt{\lambda^2-K^2}\,|y-y_0|}\right)\right\}\operatorname{sign}(y-y_0)\,d\lambda,$$

$$v_B(x,y)=\frac{1}{4\pi}\int\limits_{-\infty}^{+\infty}e^{i\lambda(x-x_0)}\left\{-\sqrt{\lambda^2-k^2}\,e^{-\sqrt{\lambda^2-k^2}\,|y-y_0|}\right.$$

$$\left.+\frac{\lambda^2}{\sqrt{\lambda^2-K^2}}\,e^{-\sqrt{\lambda^2-K^2}\,|y-y_0|}\right\}d\lambda,$$

$$w_B(x,y)=0. \qquad (1.21)$$

We also give the solution for problem C:

$$w_C(x, y) = \frac{1}{4\pi} \int\limits_{-\infty}^{+\infty} \frac{K^2}{\sqrt{\lambda^2 - K^2}} e^{i\lambda(x - x_0) - \sqrt{\lambda^2 - K^2}|y - y_0|} d\lambda,$$

$$u_C(x, y) = v_C(x, y) = 0. \tag{1.22}$$

§2. Incidence of a cylindrical wave on a rigid screen

Below we will describe the diffraction field which arises when the cylindrical wave described in §1 is incident on a rigid semi-infinite screen. The diffraction problem has the following mathematical formulation. We are required to solve the Lamé equation (1.2), which is now assumed to hold for the entire (x,y) plane except for the negative part of the OX axis where the screen is located. We have the following boundary condition on the screen:

$$\mathbf{u}(x, 0) = 0 \ (x < 0). \tag{2.1}$$

As before, the solution $\mathbf{u}(x,y)$ is sought in the class of functions which decrease exponentially at infinity (we recall that $\operatorname{Im}(k, K) > 0$), and is assumed to be continuous over the entire (x,y) plane, including the coordinate origin.

The field of the incident wave found in §1 (formulas (1.4) and (1.20)–(1.22)) will be denoted by $\mathbf{u}_0$. We introduce the diffraction field $\mathbf{U} = (U, V, W)$, isolating the incident wave $\mathbf{u}_0$ from the total field:

$$\mathbf{U} = \mathbf{u} - \mathbf{u}_0. \tag{2.2}$$

For the diffraction field we have the homogeneous Lamé equation

$$\mathbf{U} + \frac{1}{k^2} \nabla \operatorname{div} \mathbf{U} - \frac{1}{K^2} \operatorname{rot} \operatorname{rot} \mathbf{U} = 0 \tag{2.3}$$

and the nonhomogeneous boundary condition

$$\mathbf{U}(x, 0) = -\mathbf{u}_0(x, 0) \ (x < 0). \tag{2.4}$$

As before, the solutions of problems A, B and C may be obtained independently; the general solution may be expressed by the formula

$$\mathbf{U} = F_x \cdot \mathbf{U}_A + F_y \cdot \mathbf{U}_B + F_z \cdot \mathbf{U}_C. \tag{2.5}$$

Here the solution of problem C (the Sommerfeld problem) is well known. Below we will deal with problem A (the solution for B is similar). The subscript A will be omitted in formulas (2.6)–(2.35).

Our problem is to find the solution $\mathbf{U} = (U, V)$ of the equations

$$\begin{aligned} PU+QV&=0,\\ QU+RV&=0 \end{aligned} \tag{2.6}$$

for the boundary conditions

$$(x<0)\qquad \begin{aligned} U(x,\ 0)&=-u_0(x,\ 0),\\ V(x,\ 0)&=-v_0(x,\ 0). \end{aligned} \tag{2.7}$$

Here P, Q and R are defined by (1.6), while for $\mathbf{u}_0=(u_0, v_0)$ we have the integral representation (1.20). In the desired solution we isolate the odd and even parts relative to the variable y:

$$\begin{aligned} U_\pm(x,\ y)&=\frac{1}{2}[U(x,\ y)\pm U(x,\ -y)],\\ V_\pm(x,\ y)&=\frac{1}{2}[V(x,\ y)\mp V(x,\ -y)]. \end{aligned} \tag{2.8}$$

Here

$$\mathbf{U}=\mathbf{U}_+ +\mathbf{U}_-. \tag{2.9}$$

Hence for the (x,y) plane the problem breaks down into two independent problems, each of which may be formulated for the upper halfplane only. In both cases we are required to find the solution of (2.6) for $y>0$ which is continuous right up to the OX axis and which satisfies the boundary conditions

$$\begin{aligned} &U_+(x,\ 0)=-u_0(x,\ 0)\quad (x<0),\\ &\frac{\partial U_+(x,\ 0)}{\partial y}=0\quad (x>0),\\ &V_+(x,\ 0)=0\quad (-\infty<x<+\infty) \end{aligned} \tag{2.10}$$

and

$$\begin{aligned} &V_-(x,\ 0)=-v_0(x,\ 0)\quad (x<0),\\ &\frac{\partial V_-(x,\ 0)}{\partial y}=0\quad (x>0),\\ &U_-(x,\ 0)=0\quad (-\infty<x<\infty) \end{aligned} \tag{2.11}$$

respectively.

We will confine ourselves to the problem for $\mathbf{U}_+$. For brevity we omit the plus subscript in (2.12)–(2.31).

The solution (U, V) will be sought in the form

$$U(x,\ y)=\frac{1}{2\pi}\int_{-\infty}^{+\infty}-i\lambda\Phi_1(\lambda)\,e^{i\lambda x-\sqrt{\lambda^2-k^2}\,y}\,d\lambda$$

$$+\frac{1}{2\pi}\int_{-\infty}^{+\infty}\sqrt{\lambda^2-K^2}\,\Phi_2(\lambda)\,e^{i\lambda x-\sqrt{\lambda^2-K^2}\,y}\,d\lambda,$$

$$V(x,\ y)=\frac{1}{2\pi}\int_{-\infty}^{+\infty}\sqrt{\lambda^2-k^2}\,\Phi_1(\lambda)\,e^{i\lambda x-\sqrt{\lambda^2-k^2}\,y}\,d\lambda$$

$$+\frac{1}{2\pi}\int_{-\infty}^{+\infty}i\lambda\Phi_2(\lambda)\,e^{i\lambda x-\sqrt{\lambda^2-K^2}\,y}\,d\lambda, \tag{2.12}$$

where $\Phi_1(\lambda)$ and $\Phi_2(\lambda)$ are assumed to be independent.

Thus for a very broad class of functions $\Phi_i(\lambda)$ the functions $U(x,y)$ and $V(x,y)$ automatically satisfy (2.6) and have the requisite behavior at infinity. The branches at the radicals are chosen in the same way as in §1.

The boundary condition (2.10) for V yields the expression

$$\frac{1}{2\pi}\int_{-\infty}^{+\infty}\left[\sqrt{\lambda^2-k^2}\,\Phi_1(\lambda)+i\lambda\Phi_2(\lambda)\right]e^{i\lambda x}d\lambda=0, \tag{2.13}$$

which holds for all x. We can satisfy this expression by setting

$$\Phi_1(\lambda)=-i\lambda\Psi(\lambda),\quad \Phi_2(\lambda)=\sqrt{\lambda^2-k^2}\,\Psi(\lambda). \tag{2.14}$$

Here $\Psi(\lambda)$ is a new unknown function. To ensure that $U(x,y)$ is continuous at the coordinate origin, we require that $\Psi(\lambda)$ have the following behavior at infinity:

$$\Psi(\lambda)=0(|\lambda|^{-1-\varepsilon})\quad(\varepsilon>0). \qquad |\lambda|\to\infty \tag{2.15}$$

Using the boundary conditions for U, we obtain for $\Psi(\lambda)$ the pair of integral equations

$$\int_{-\infty}^{+\infty}\sqrt{\lambda^2-k^2}\,\Psi(\lambda)\,e^{i\lambda x}d\lambda=0\quad(x>0),$$

$$\int_{-\infty}^{+\infty}\Big[\left(-\lambda^2+\sqrt{\lambda^2-k^2}\sqrt{\lambda^2-K^2}\right)\Psi(\lambda)+\frac{1}{2}e^{-i\lambda x_0}$$

$$\times\left(\frac{\lambda^2}{\sqrt{\lambda^2-k^2}}e^{-\sqrt{\lambda^2-k^2}\,|y_0|}-\sqrt{\lambda^2-K^2}\,e^{-\sqrt{\lambda^2-K^2}\,|y_0|}\right)\Big]$$

$$\times e^{i\lambda x}d\lambda=0\quad(x<0). \tag{2.16}$$

For the first of these to hold, we set

$$\sqrt{\lambda^2-k^2}\,\Psi(\lambda)=F^+(\lambda), \tag{2.17}$$

where the function $F^+(\lambda)$ is analytic in the upper halfplane of the complex plane λ. The second expression in (2.16) similarly leads to

$$(-\lambda^2+\sqrt{\lambda^2-k^2}\sqrt{\lambda^2-K^2})\,\Psi(\lambda)+\frac{1}{2}e^{-i\lambda x_0}$$
$$\times\left(\frac{\lambda^2}{\sqrt{\lambda^2-k^2}}e^{-\sqrt{\lambda^2-k^2}\,|y_0|}-\sqrt{\lambda^2-K^2}\,e^{-\sqrt{\lambda^2-K^2}\,|y_0|}\right)=F^-(\lambda), \tag{2.18}$$

where $F^-(\lambda)$ is analytic in the lower halfplane.

From (2.17) and (2.18) we obtain the following Riemann boundary value problem [7]: to find two functions, $F^+(\lambda)$ which is analytic in the upper halfplane and $F^-(\lambda)$ which is analytic in the lower, which satisfy the expression

$$(\lambda^2-\sqrt{\lambda^2-k^2}\sqrt{\lambda^2-K^2})\frac{F^+(\lambda)}{\sqrt{\lambda+k}}+\sqrt{\lambda-k}\,F^-(\lambda)=\frac{1}{2}e^{-i\lambda x_0}\cdot\sqrt{\lambda-k}$$
$$\times\left(-\sqrt{\lambda^2-K^2}\,e^{-\sqrt{\lambda^2-K^2}\,|y_0|}+\frac{\lambda^2}{\sqrt{\lambda^2-k^2}}e^{-\sqrt{\lambda^2-k^2}\,|y_0|}\right) \tag{2.19}$$

on the real axis and which at infinity are of order

$$F^+(\lambda)\underset{|\lambda|\to\infty}{=}0(|\lambda|^{-\varepsilon}),\quad F^-(\lambda)\underset{|\lambda|\to\infty}{=}0(|\lambda|^{-1-\varepsilon}). \tag{2.20}$$

To solve our boundary value problem we must first represent the function

$$l(\lambda)=\lambda^2-\sqrt{\lambda^2-k^2}\sqrt{\lambda^2-K^2} \tag{2.21}$$

as the product of two functions, one of which $[l^+(\lambda)]$ is analytic in the upper halfplane and the other $[l^-(\lambda)]$ in the lower:

$$l(\lambda)=l^+(\lambda)\cdot l^-(\lambda). \tag{2.22}$$

This representation (called factorization) results from the use of the Cauchy integrals [3, 6]. The function $l(\lambda)$ is defined on a two-sheeted Riemann surface and has four branch points ($l=\pm k,\pm K$) of multiplicity two on it. Joining the points $\lambda=k$ and $\lambda=K$ as well as the points $\lambda=-k$ and $\lambda=-K$ by a straight-line cut, we can isolate the sheets of this surface. The principal sheet, as before, is fixed by the requirement (1.9). The function $l(\lambda)$ has no roots on the principal sheet (two of its roots $\lambda=\pm kK/\sqrt{(k^2+K^2)}$ are located on the additional sheet) and is bounded as $|\lambda|\to\infty$:

$$\lim_{|\lambda|\to\infty} l(\lambda)=\frac{k^2+K^2}{2}\equiv a^2. \tag{2.23}$$

Hence on the principal sheet the function $\ln(l(\lambda)/a^2)$ can be uniquely determined, where

$$\lim_{|\lambda|\to\infty} \ln \frac{l(\lambda)}{a^2} = 0. \tag{2.24}$$

By the Cauchy theorem,

$$\ln \frac{l(\lambda)}{a^2} = \frac{1}{2\pi i} \int_C \frac{\ln \frac{l(\tau)}{a^2}}{\tau - \lambda} d\tau, \tag{2.25}$$

where C is an arbitrary closed contour which clears the cuts and traverses the point $\tau = \lambda$ in the positive direction. Allowing for (2.24), we have

$$\frac{1}{2\pi i} \int_C \frac{\ln \frac{l(\tau)}{a^2}}{\tau - \lambda} d\tau = \frac{1}{2\pi i} \int_{\Gamma_+} \frac{\ln \frac{l(\tau)}{a^2}}{\tau - \lambda} d\tau + \frac{1}{2\pi i} \int_{\Gamma_-} \frac{\ln \frac{l(\tau)}{a_1}}{\tau - \lambda} d\tau. \tag{2.26}$$

Here Γ_+ is a contour which goes (in the clockwise direction) around the cut joining the points $\tau = -k$ and $\tau = -K$. The contour Γ^- similarly traverses the cut between $\tau = k$ and $\tau = K$. Introducing the functions

$$l^{\pm}(\lambda) = a \cdot \exp\left[\frac{1}{2\pi i} \int_{\Gamma_\pm} \frac{\ln \frac{l(\tau)}{a^2}}{\tau - \lambda} d\tau\right], \tag{2.27}$$

we arrive at the required representation (2.22), since each of the integrals in (2.27) specifies a function which is analytic outside the contour of integration. It is clear from (2.27) that $l^{\pm}(\lambda)$ are bounded outside their integration contours and

$$\lim_{|\lambda|\to\infty} l^{\pm}(\lambda) = a. \tag{2.28}$$

Allowing for (2.22), (2.19) transforms to

$$\frac{l^+(\lambda) \cdot F^+(\lambda)}{\sqrt{\lambda + k}} - X_1^+(\lambda) = -\frac{F(\lambda)\sqrt{\lambda - k}}{l^-(\lambda)} - X_1^-(\lambda), \tag{2.29}$$

where

$$X_1(\lambda) = \frac{1}{2\pi i} \int_{-\infty}^{+\infty} \frac{\sqrt{\tau - k}}{2l^-(\tau)} e^{-i\tau x_0} \cdot \left\{-\sqrt{\tau^2 - K^2}\, e^{-\sqrt{\tau^2 - K^2}\,|y_0|} \right.$$

$$\left. + \frac{\tau^2}{\sqrt{\tau^2 - k^2}} e^{-\sqrt{\tau^2 - k^2}\,|y_0|}\right\} \cdot \frac{d\tau}{\tau - \lambda} = \begin{cases} X_1^+(\lambda) & \text{for } \operatorname{Im}\lambda > 0, \\ X_1^-(\lambda) & \text{for } \operatorname{Im}\lambda < 0 \end{cases} \tag{2.30}$$

and the Sohockiĭ formulas [7] have been taken into account.

According to Riemann's theorem on analytic continuation through a contour, the left and right sides of (2.29) specify a single function which is analytic over the entire complex plane λ; according to Liouville's theorem (allowing for the behavior of this function at infinity), this function is identically zero.

Now (2.14), (2.17) and (2.29) yield

$$\Phi_1(\lambda) = -\frac{i\lambda}{l^+(\lambda)\sqrt{\lambda-k}}\,X_1^+(\lambda), \quad \Phi_2(\lambda) = \frac{\sqrt{\lambda+k}}{l^+(\lambda)}\,X_1^+(\lambda). \tag{2.31}$$

Representations (2.12) yield the solution for problem A_+ :

$$U_+(x, y) = \frac{1}{2\pi}\int_{-\infty}^{+\infty} \frac{\chi_1^+(\lambda)}{\sqrt{\lambda-k}\,l^+(\lambda)}\left[-\lambda^2 e^{-\sqrt{\lambda^2-k^2}|y|} + \sqrt{\lambda^2-k^2}\sqrt{\lambda^2-K^2}\,e^{-\sqrt{\lambda^2-K^2}|y|}\right] e^{i\lambda x}\,d\lambda,$$

$$V_+(x, y) = \frac{1}{2\pi}\int_{-\infty}^{+\infty} \frac{i\lambda\sqrt{\lambda+k}\,\chi_1^+(\lambda)}{l^+(\lambda)} \times \left[-e^{-\sqrt{\lambda^2-k^2}|y|} + e^{-\sqrt{\lambda^2-K^2}|y|}\right] e^{i\lambda x}\,\mathrm{sign}\,y\,d\lambda. \tag{2.32}$$

Similarly we find the solution for A_- :

$$U_-(x, y) = \frac{1}{2\pi}\int_{-\infty}^{+\infty} \frac{i\lambda\sqrt{\lambda+K}\,\chi_2^+(\lambda)}{l^+(\lambda)}\left[-e^{-\sqrt{\lambda^2-k^2}|y|} + e^{-\sqrt{\lambda^2-K^2}|y|}\right] e^{i\lambda x}\,\mathrm{sign}\,y\,d\lambda,$$

$$V_-(x, y) = \frac{1}{2\pi}\int_{-\infty}^{+\infty} \frac{\chi_2^+(\lambda)}{\sqrt{\lambda-K}\,l^+(\lambda)} \times \left[-\lambda^2 e^{-\sqrt{\lambda^2-K^2}|y|} + \sqrt{\lambda^2-k^2}\sqrt{\lambda^2-K^2}\,e^{-\sqrt{\lambda^2-k^2}|y|}\right] e^{i\lambda x}\,d\lambda, \tag{2.33}$$

where

$$X_2(\lambda) = \frac{1}{2\pi i}\int_{-\infty}^{+\infty} \frac{i\sqrt{\tau-K}}{2l^-(\tau)}\,e^{-i\tau x_0}\,\mathrm{sign}\,y_0 \times \left[e^{-\sqrt{\tau^2-K^2}|y_0|} - e^{-\sqrt{\tau^2-k^2}|y_0|}\right]\frac{d\tau}{\tau-\lambda}. \tag{2.34}$$

Our final expressions for the solution of problem A in the presence of a rigid screen are

$$u(x, y) = u_0(x, y) + U_+(x, y) + U_-(x, y),$$
$$v(x, y) = v_0(x, y) + V_+(x, y) + V_-(x, y),$$
$$w(x, y) = 0. \tag{2.35}$$

Problem B is solved similarly. The solution is obtained from the preceding formulas, but the expressions for $X_1(\lambda)$ and $X_2(\lambda)$ change:

$$X_1(\lambda)=\frac{1}{2\pi i}\int_{-\infty}^{+\infty}\frac{i\sqrt{\tau-k}}{2l^-(\tau)}e^{-i\tau x_0}$$

$$\times\left[e^{-\sqrt{\tau^2-K^2}|y_0|}-e^{-\sqrt{\tau^2-k^2}|y_0|}\right]\operatorname{sign} y_0\cdot\frac{d\tau}{\tau-\lambda}, \qquad (2.36)$$

$$X_2(\lambda)=\frac{1}{2\pi i}\int_{-\infty}^{+\infty}\frac{\sqrt{\tau-K}}{2l^-(\tau)}e^{-i\tau x_0}$$

$$\times\left[-\sqrt{\tau^2-k^2}\,e^{-\sqrt{\tau^2-k^2}|y_0|}+\frac{\tau^2}{\sqrt{\tau^2-K^2}}e^{-\sqrt{\tau^2-K^2}|y_0|}\right]\frac{d\tau}{\tau-\lambda}. \qquad (2.37)$$

We also give the solution for problem C:

$$W(x,\ y)=W_0(x,\ y)+\frac{1}{2\pi}\int_{-\infty}^{+\infty}\frac{\chi_3^+(\lambda)}{\sqrt{\lambda-K}}e^{i\lambda x-\sqrt{\lambda^2-K^2}|y|}d\lambda, \qquad (2.38)$$

where

$$X_3(\lambda)=\frac{1}{2\pi i}\int_{-\infty}^{+\infty}\frac{-K^2}{2\sqrt{\tau+K}}e^{-i\tau x_0-\sqrt{\tau^2-K^2}|y_0|}\frac{d\tau}{\tau-\lambda}. \qquad (2.39)$$

§3. Incidence of a cylindrical wave on a crack

Below we will construct a solution for (1.5) which satisfies the boundary conditions $\sigma_{yy}=\sigma_{xy}=0$ for $x<0$, $y=0$ and the conditions at infinity. Thus we will solve problem A in the presence of a semi-infinite crack. Problems B and C are solved similarly. As before, $w_A=0$, and therefore the third boundary condition $\sigma_{xy}=0$ in problem A need not be given. The subscript A is omitted in (3.1)–(3.19).

We will formulate the boundary conditions in terms of displacements:

$$\sigma_{yy}=\frac{E}{(1+\sigma)(1-2\sigma)}\left[(1-\sigma)\frac{\partial v}{\partial y}+\sigma\frac{\partial u}{\partial x}\right]=0,$$

$$\sigma_{xy}=\frac{E}{2(1+\sigma)}\left(\frac{\partial u}{\partial y}+\frac{\partial v}{\partial x}\right)=0,\quad (y=0,\ x<0). \qquad (3.1)$$

It is easy to show that

$$\frac{\sigma}{1-\sigma}=\frac{K^2-2k^2}{K^2},$$

so that the boundary conditions assume the final form

$$\frac{\partial v}{\partial y}+\frac{K^2-2k^2}{K^2}\frac{\partial u}{\partial x}=0,$$

$$\frac{\partial u}{\partial y}+\frac{\partial v}{\partial x}=0, \quad (x<0,\ y=0). \tag{3.2}$$

As was done in §2, we isolate from the total field **u** the diffraction field **U** in accordance with (2.2). We represent it as the sum of two terms $\mathbf{U}_+$ and $\mathbf{U}_-$ (formulas (2.8) and (2.9)), each of which may be considered in the upper halfplane only. As before, we obtain for $\mathbf{U}_+$ and $\mathbf{U}_-$ two independent problems, of which we will henceforth consider only the problem for $\mathbf{U}_+$. The subscript plus sign will be omitted.

We formulate the boundary conditions for $\mathbf{U}_+$:

$$\frac{\partial V}{\partial y}+\frac{K^2-2k^2}{K^2}\frac{\partial U}{\partial x}=-\left(\frac{\partial v_0}{\partial y}+\frac{K^2-2k^2}{K^2}\frac{\partial u_0}{\partial x}\right) \quad (x<0,\ y=0),$$

$$\frac{\partial U}{\partial y}+\frac{\partial V}{\partial x}=0 \quad (x<0,\ y=0),$$

$$\frac{\partial U}{\partial y}=0, \quad V=0 \quad (x>0,\ y=0). \tag{3.3}$$

Here **U** is assumed to be continuous over the entire (x,y) plane with a cut along the negative part of the OX axis right up to the cut itself.

As before, the solution is sought in the form (2.12). It is clear from (3.3) that the boundary condition $\partial U/\partial y+\partial V/\partial x=0$ holds for the entire OX axis, whence

$$\int_{-\infty}^{+\infty}\left[2i\lambda\sqrt{\lambda^2-k^2}\,\Phi_1(\lambda)-(2\lambda^2-K^2)\,\Phi_2(\lambda)\right]e^{i\lambda x}d\lambda=0 \quad (-\infty<x<\infty). \tag{3.4}$$

To satisfy (3.4), we set

$$\Phi_1(\lambda)=(2\lambda^2-K^2)\,\Psi(\lambda), \quad \Phi_2(\lambda)=2i\lambda\sqrt{\lambda^2-k^2}\,\Psi(\lambda). \tag{3.5}$$

Boundary conditions (3.3) now generate the following pair of integral equations:

$$\int_{-\infty}^{+\infty}\sqrt{\lambda^2-k^2}\,\Psi(\lambda)\,e^{i\lambda x}\,d\lambda=0 \quad (x>0),$$

$$\int_{-\infty}^{+\infty}l_1(\lambda)\,\Psi(\lambda)\,e^{i\lambda x}d\lambda=\int_{-\infty}^{+\infty}f_1(\lambda)\,e^{i\lambda x}d\lambda \ (x<0), \tag{3.6}$$

where

$$l_1(\lambda)=4\lambda^2\sqrt{\lambda^2-k^2}\sqrt{\lambda^2-K^2}-(K^2-2\lambda^2)^2, \tag{3.7}$$

$$f_1(\lambda)=\frac{i\lambda}{2\sqrt{\lambda^2-k^2}}\,e^{-i\lambda x_0}\left\{2\sqrt{\lambda^2-k^2}\sqrt{\lambda^2-K^2}\,e^{-\sqrt{\lambda^2-K^2}|y_0|}\right.$$

$$\left.+(K^2-2\lambda^2)\,e^{-\sqrt{\lambda^2-k^2}|y_0|}\right\}. \tag{3.8}$$

Solving (3.6), we obtain

$$\sqrt{\lambda^2 - k^2}\, \psi(\lambda) = F^+(\lambda),$$
$$l_1(\lambda)\, \psi(\lambda) = f_1(\lambda) + F^-(\lambda), \tag{3.9}$$

where, as before, the + (−) superscript denotes analyticity in the upper (lower) halfplane. Eliminating $\psi(\lambda)$, we obtain the Riemann boundary value problem

$$l_1(\lambda) \frac{F^+(\lambda)}{\sqrt{\lambda + k}} - \sqrt{\lambda - k}\, F^-(\lambda) = \sqrt{\lambda - k}\, f_1(\lambda). \tag{3.10}$$

The factorization of the factor $l_1(\lambda)$ is given in [3]. It is done in the same way as the factorization of $l(\lambda)$ in §2. The only difference is that $l_1(\lambda)$ behaves at infinity like $O(|\lambda|^2)$ on the principal sheet, and has the pair of roots $\lambda = \pm\lambda_R$, where λ_R is the wave number of the Rayleigh wave. In factorizing, therefore, we should first isolate from $l_1(\lambda)$ the factor $(\lambda^2 - \lambda_R^2)$:

$$l_1^{\pm}(\lambda) = (\lambda \pm \lambda_R) \sqrt{2(K^2 - k^2)} \times \exp\left[\frac{1}{2\pi i} \int_{\Gamma_\pm} \frac{\ln \dfrac{l_1(\tau)}{(\tau^2 - \lambda_R^2)\, 2(K^2 - k^2)}}{\tau - \lambda}\, d\tau\right]. \tag{3.11}$$

As a result we have

$$l_1^+(\lambda) \frac{F^+(\lambda)}{\sqrt{\lambda + k}} - X_1^+(\lambda) = \frac{\sqrt{\lambda - k}}{l_1^-(\lambda)} F^-(\lambda) - X_1^-(\lambda), \tag{3.12}$$

where

$$X_1(\lambda) = \frac{1}{2\pi i} \int_{-\infty}^{+\infty} \frac{\sqrt{\tau - k}}{l_1^-(\tau)} f_1(\tau) \frac{d\tau}{\tau - \lambda}. \tag{3.13}$$

Finally,

$$U_+(x, y) = \frac{i}{2\pi} \int_{-\infty}^{+\infty} \frac{\lambda \chi_1^+(\lambda)}{\sqrt{\lambda - k}\, l_1^+(\lambda)} \times \left\{2\sqrt{\lambda^2 - k^2} \sqrt{\lambda^2 - K^2}\, e^{-\sqrt{\lambda^2 - K^2}|y|} + (K^2 - 2\lambda^2)\, e^{-\sqrt{\lambda^2 - k^2}|y|}\right\} e^{i\lambda x} d\lambda, \tag{3.14}$$

$$V_+(x, y) = \frac{1}{2\pi} \int_{-\infty}^{+\infty} \frac{\sqrt{\lambda + k}\, \chi_1^+(\lambda)}{l_1^+(\lambda)} \times \left\{(2\lambda^2 - K^2)\, e^{-\sqrt{\lambda^2 - k^2}|y|} - 2\lambda^2 e^{-\sqrt{\lambda^2 - K^2}|y|}\right\} \operatorname{sign} y \cdot e^{i\lambda x} d\lambda. \tag{3.15}$$

Problem A_- is solved similarly:

$$U_{-}(x, y) = \frac{1}{2\pi}\int_{-\infty}^{+\infty} \frac{\sqrt{\lambda+K}\,\chi_2^{+}(\lambda)}{l_1^{+}(\lambda)}$$

$$\times\left\{(K^2-2\lambda^2)\,e^{-\sqrt{\lambda^2-K^2}|y|}+2\lambda^2 e^{-\sqrt{\lambda^2-k^2}|y|}\right\}\operatorname{sign} y\cdot e^{i\lambda x}d\lambda, \tag{3.16}$$

$$V_{-}(x, y) = \frac{i}{2\pi}\int_{-\infty}^{+\infty} \frac{\lambda\cdot\chi_2^{+}(\lambda)}{\sqrt{\lambda-K}\,l_1^{+}(\lambda)}$$

$$\times\left\{2\sqrt{\lambda^2-k^2}\sqrt{\lambda^2-K^2}\,e^{-\sqrt{\lambda^2-k^2}|y|}+(K^2-2\lambda^2)\,e^{-\sqrt{\lambda^2-K^2}|y|}\right\}e^{i\lambda x}d\lambda, \tag{3.17}$$

$$\mathrm{X}_2(\lambda) = \frac{1}{2\pi i}\int_{-\infty}^{+\infty} \frac{\sqrt{\tau-K}}{l_1^{-}(\tau)}\,f_2(\tau)\,\frac{d\tau}{\tau-\lambda}, \tag{3.18}$$

$$f_2(\lambda) = \frac{1}{2}\,e^{-i\lambda x_0}\cdot\operatorname{sign} y_0\left\{(2\lambda^2-K^2)\,e^{-\sqrt{\lambda^2-K^2}|y_0|}-2\lambda^2 e^{-\sqrt{\lambda^2-k^2}|y_0|}\right\}. \tag{3.19}$$

We will now give the solutions for problems B and C: the solution of B is obtained from (3.13)–(3.18), where we must set

$$f_1(\lambda) = \frac{1}{2}\,e^{-i\lambda x_0}\cdot\operatorname{sign} y_0\left\{(2\lambda^2-K^2)\,e^{-\sqrt{\lambda^2-K^2}|y_0|}-2\lambda^2 e^{-\sqrt{\lambda^2-k^2}|y_0|}\right\}; \tag{3.20}$$

$$f_2(\lambda) = \frac{1}{2}\cdot\frac{i\lambda}{\sqrt{\lambda^2-K^2}}\,e^{-i\lambda x_0}\left\{-2\sqrt{\lambda^2-k^2}\sqrt{\lambda^2-K^2}\,e^{-\sqrt{\lambda^2-k^2}|y_0|}\right.$$

$$\left.+(2\lambda^2-K^2)\,e^{-\sqrt{\lambda^2-K^2}|y_0|}\right\}. \tag{3.21}$$

The solution of problem C is

$$w(x, y) = w_0(x, y) + \frac{1}{2\pi}\int_{-\infty}^{+\infty} \frac{\chi^{+}(\lambda)}{\sqrt{\lambda+K}}\,e^{i\lambda x-\sqrt{\lambda^2-K^2}|y|}\,d\lambda, \tag{3.22}$$

where

$$\mathrm{X}(\lambda) = \frac{1}{2\pi i}\int_{-\infty}^{+\infty} \frac{K^2}{2\sqrt{\tau-K}}\,e^{-i\tau x_0-\sqrt{\tau^2-K^2}|y_0|}\,\frac{d\tau}{\tau-\lambda}\operatorname{sign} y_0. \tag{3.23}$$

Problems of diffraction of a purely longitudinal or purely transverse cylindrical wave are also of interest. It is usual to formulate problems of this type in terms of scalar and vector potentials. In the first case we have

$$\Delta\varphi + k^2\varphi = -\delta(x-x_0,\ y-y_0),$$
$$\Delta\psi + K^2\psi = 0, \tag{3.24}$$

where

$$u = \frac{\partial\varphi}{\partial x} + \frac{\partial\psi}{\partial y},$$
$$v = \frac{\partial\varphi}{\partial y} - \frac{\partial\psi}{\partial x}\quad (w=0). \tag{3.25}$$

The Lamé equations for the case (3.24), (3.25) have the form

$$\left.\begin{aligned} Pu+Qv&=-\frac{1}{k^2}\frac{\partial}{\partial x}\delta(x-x_0,\ y-y_0)=-\frac{1}{k^2}\frac{\partial}{\partial x_0}\delta(x-x_0,\ y-y_0),\\ Qu+Rv&=-\frac{1}{k^2}\frac{\partial}{\partial y}\delta(x-x_0,\ y-y_0)=-\frac{1}{k^2}\frac{\partial}{\partial y_0}\delta(x-x_0,\ y-y_0).\end{aligned}\right\} \quad (3.26)$$

As a result, for any of the screens under consideration, the solution of the problem of the diffraction of a transverse cylindrical wave is specified by the formula

$$\mathbf{u}=\frac{1}{k^2}\left(\frac{\partial\mathbf{u}_A}{\partial x_0}+\frac{\partial\mathbf{u}_B}{\partial y_0}\right). \quad (3.27)$$

The diffraction field of a longitudinal cylindrical wave is similarly specified:

$$\mathbf{u}=\frac{1}{K^2}\left(\frac{\partial\mathbf{u}_A}{\partial y_0}+\frac{\partial\mathbf{u}_B}{\partial x_0}\right). \quad (3.28)$$

Bibliography

1. M.M. Fridman, *Diffraction of a plane elastic wave relative to a semi-infinite straight rigidly attached screen*, Uč. Zap. Leningrad. Univ. Ser. Mat. Meh. 114 (1949). no. 17. (Russian)
2. ———, *The diffraction of a plane elastic wave by a semi-infinite rectilinear slit, free of stress*, Dokl. Akad. Nauk SSSR 66 (1949), 21–24. (Russian) MR 10, 763.
3. A.W. Maue, *Die Beugung elastischer Wellen an der Halbebene*, Z. Angew. Math. Mech. **33** (1953), 1–10. MR **14,** 929.
4. A.F. Filippov, *Certain problems of diffraction of plane elastic waves*, Prikl. Mat. Meh. **20** (1956), 688–703. (Russian) MR **19,** 196.
5. ———, *A three-dimensional problem of diffraction of an elastic wave at a sharp edge*, Prikl. Mat. Meh. **23** (1959), 691–696 = J. Appl. Math. Mech. **23** (1959), 989–996. MR **22** #3248.
6. D.P. Kouzov, *Diffraction of a cylindrical hydro-acoustic wave at the junction of two semi-infinite plates*, Prikl. Mat. Meh. **33** (1969), 240–250 = J. Appl. Math. Mech. **33** (1969), 225–234.
7. F.D. Gahov, *Boundary-value problems*, Fizmatgiz, Moscow, 1958; English transl., Pergamon Press, Oxford; Addison-Wesley, Reading, Mass., 1966. MR **21** #2879; **33** #6311.

Translated by:
M. Lopez-Morillas

ON A LOCAL METHOD OF SOLUTION OF A NONSTATIONARY INVERSE PROBLEM FOR A NONHOMOGENEOUS STRING

A. S. BLAGOVEŠČENSKIĬ

UDC 517 944/947

Abstract. The paper is devoted to the determination of an unknown coefficient in the problem of the vibration of an inhomogeneous string. This problem was solved by M.G. Kreĭn, who reduced it to a one-dimensional stationary problem. The author has given a direct, elementary solution of this problem. M.G. Kreĭn's problem is solved in this paper by means of an integral equation which is similar to a Volterra equation.

Bibliography: 5 items.

1. The goal of this paper is the following: to give a new proof of the fundamental results of M.G. Kreĭn in the theory of inverse boundary problems for the equation of a string [1–3]. We would like our proof to have the following attributes: 1) the basic idea should be simple; 2) it should be local (this requirement could be formulated differently: the proof should be nonstationary). In order to clarify the meaning of the second requirement, we formulate the problem to be studied. We wish to find the function $c(x) > 0$ in the equation

$$u_{tt}^{+} = Lu^{+}, \quad x > 0, \tag{1}$$

where

$$Lu^{+} \equiv u_{xx}^{+} - \frac{c'(x)}{c(x)} u_{x}^{+} = c(x) \frac{\partial}{\partial x} \left(\frac{u_{x}^{+}}{c(x)} \right), \tag{1a}$$

if it is known that

$$u^{+}|_{x=0} = f^{+}(t), \; u_{x}^{+}|_{x=0} = c(0)\delta(t), \; u^{+}|_{t<0} \equiv 0. \tag{I}$$

This problem for equation (1), with data (I), will be called problem I. Suppose that the function $c(x)$ is given; then, as is well known, problem I is uniquely solvable if one omits the first of the conditions I (let us call this the direct problem), and then the values of $u^{+}(x,t)$ for $x = 0$ are obtained as a by-product of the solution of the direct problem. Also, inasmuch as equation (1) describes the propagation of waves which spread with unit velocity, the values of $u^{+}(0,t)$ for $t \in (0, 2T)$ depend only on the behavior of $c(x)$ on the interval $(0, T)$. Therefore one should naturally expect the converse to be true: that the function $c(x)$ depends, on each interval $(0, T)$, only on the behavior of $f^{+}(t)$ on the interval $(0, 2T)$. It will be seen that the dependence of $c(x)$ on $f^{+}(t)$ is indeed local in this sense, and, clearly, this property will be reflected in the results which we will prove. However, the approaches which were

---

AMS 1970 *subject classifications*. Primary 35L20, 35R30; Secondary 45D05.

previously employed in the study of similar problems failed to exhibit this property in a clear manner. Indeed, the usual approach, which migh be called spectral or stationary, consists in taking the Fourier (or Laplace) transform with respect to the time variable, and then studying the problem of reconstructing $c(x)$ by considering properties of the eigenfunctions arising from this spectral problem. Clearly the behavior of the eigenfunctions depends on the behavior of $c(x)$ on the entire halfline $0 < x < \infty$. It should be remarked that the method of proof we propose is related to that employed by B.S. Pariĭskiĭ [4], but in his interesting paper a) the problem studied is a different one, b) the proof is not completed, and c) there are errors. The basic considerations in [4] are carried out in the region $t > x$; in the present paper we work in the region $|t| < x$.

2. We remark that problem I is equivalent to the problem

$$\left.\begin{aligned} u^+_{tt} &= c^2(z)\, u^+_{zz}, \quad z > 0, \\ u^+|_{z=0} = f^+(t), \quad & u^+_z|_{z=0} = \delta(t), \quad u^+|_{t<0} \equiv 0, \end{aligned}\right\}$$

where $c(z)$ is unknown. This problem is reduced to problem I by the change of variable $x = \int_0^z (1/c(\zeta))\, d\zeta$.

3. To make the reader's orientation easier, we present a diagram of the interdependence of the proofs given later. In it we indicate what will be the principal object of our study, along with problem I: namely, the integral equation of M.G. Kreĭn (see §6), which, in view of its fundamental role, will be called equation (F) (fundamental); we will refer to the necessary and sufficient condition for the solvability of problem I as M.G. Kreĭn's condition (K) on the function $f^+(t)$ (see §8).[1]

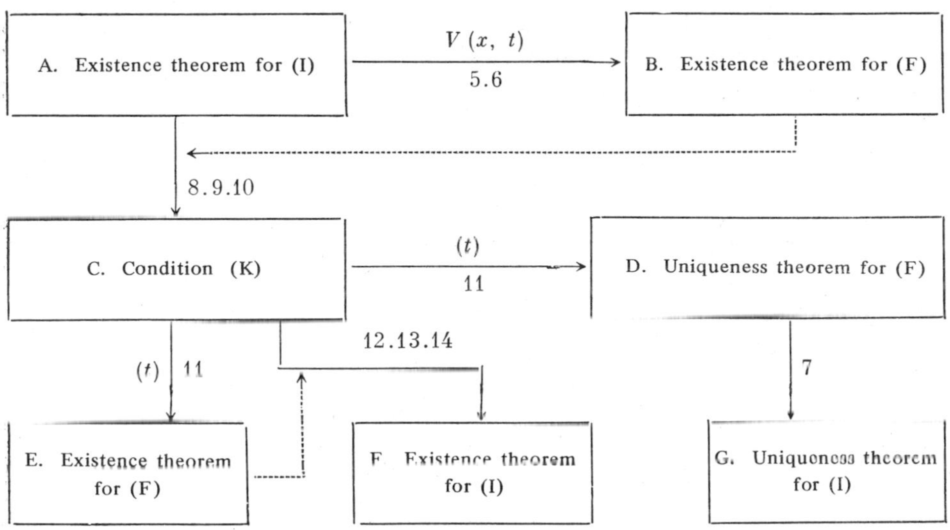

[1] The reader will most probably find it convenient in the following to return to this diagram.

We should explain the meaning of this diagram. An arrow replaces the word "implies". For instance, the arrow going from A to B (briefly: A $\to$ B) means "The solvability of problem I implies that Kreĭn's equation (F) is also solvable". The letter V above this arrow means that during the course of the proof, we introduce a new function $V(x,t)$, which plays an important role in the following (it itself is a solution of equation (F)). Dotted arrows are auxiliary. For instance, the dotted arrow leading from B to A $\to$ C means "The existence theorem for problem I implies that condition (K) holds; equation (F) is used in an essential manner in the proof, but condition (K) is not a consequence of the solvability of equation (F), if the latter has not been obtained from problem I". The symbol (t) above certain arrows means that the corresponding assertions are trivial, but that they still deserve some attention. The numbers near the arrows refer to the sections in which the corresponding assertions are proved.

The main thrust of the diagram may be formulated in the following concise manner:

Assuming condition (K), the inverse problem (I) is equivalent to equation (F). Assuming condition (K), the solution to (I) exists, and is unique.

4. Let us turn to the proofs. We prove assertion A $\to$ B (see the diagram), i.e. the statement that the solvability of problem (I) entails the existence of a function $V(x,t)$ which is a solution of Kreĭn's integral equation (which will be written out below). The function $V(x,t)$ is constructed from $c(x)$ via the solution of a certain Cauchy problem. We must formulate problem (I) more precisely. First we will assume that the desired function $c(x)$ is smooth for $x > 0$. It is then not difficult to show that a necessary condition for the solvability of problem (I) is: a) the function $f^+(t)$ is smooth for $t \geqslant 0$; b) $f^+(+0) < 0$. Below we shall always assume that $f^+(t)$ satisfies these conditions.

5. We now perform certain auxiliary constructions. Let $u^+(x,t)$, $c(x)$ be a solution-pair of problem (I). Let $u(x,t)$ be defined in terms of $u^+(x,t)$ as follows:

$$u(x,t) = u^+(x,t), \quad t \geqslant 0,$$
$$u(x,t) = -u^+(x,-t), \quad t \leqslant 0.$$

Below we shall refer to $u(x,t)$ as an odd continuation of $u^+(x,t)$ (although this term is not quite precise). Let $f(t)$ be the odd continuation of $f^+(t)$, i.e. the classical (not generalized) function defined by the equations

$$f(t) = f^+(t) \quad \text{for } t > 0,$$
$$f(t) = -f^+(-t) \quad \text{for } t < 0.$$

It is easily seen that $u(x,t)$ is a solution of the following problem:

$$\left.\begin{aligned} u_{tt} &= Lu \quad (x > 0), \\ u|_{x=0} &= f(t), \quad u_x|_{x=0} = 0. \end{aligned}\right\} \tag{II}$$

For us the fact that for $x = 0$ the second Cauchy condition on u is homogeneous will be of great importance. Being a solution of problem (I), the function $u^+(x,t)$ has the property

$$u^+(x, t) \equiv 0 \quad \text{for } t < x. \tag{2}$$

Therefore

$$u(x, t) \equiv 0 \quad \text{for } |t| < x. \tag{3}$$

Let $W(x,t)$ be the solution of the problem

$$\left.\begin{aligned} &W_{tt} = LW \quad \text{for } x > 0, \\ &W|_{x=0} = \delta(t), \quad W_x|_{x=0} = 0. \end{aligned}\right\} \tag{III}$$

Obviously $W(x,t)$ has the following properties:

$$W(x, t) \equiv 0 \quad \text{for } |t| > x, \tag{4}$$

$$W(x, t) = \frac{1}{2}\sqrt{\frac{c(x)}{c(0)}}\,[\delta(t+x) + \delta(t-x)] + \tilde{W}(x, t), \tag{5}$$

where $\tilde{W}(x,t)$ is a classical function, smooth for $x > |t|$. Besides all this, it is useful to observe that if the functions $c(x)$ and $u^+(x,t)$ are continued as even functions to negative values of x, then the function $u^+(x,t)$ (in order to avoid burdening this work with notation, we will not introduce a special symbol for it) is found to be a solution of the following problem:

$$\left.\begin{aligned} &u^+_{tt} = Lu^+, \quad -\infty < x < \infty, \\ &u^+|_{t=0} = 0, \quad u^+_t|_{t=0} = -2c(0)\,\delta(x). \end{aligned}\right\} \tag{IV}$$

6. There is an obvious connection between $u(x,t)$ and $W(x,t)$:

$$u(x, t) = \int W(x, s)\, f(t-s)\, ds. \tag{6}$$

This relation will lead us to the integral equation (F).[2]

---

[2] It is easy to get an integral equation for the inverse problem, one different from equation (F), from formulas (3)–(6). Indeed, taking $x > |t|$ in (6), we get

$$\int W(x, \tau)\, f(t-\tau)\, d\tau = 0; \tag{6a}$$

multiply (6a) by $\sqrt{c(0)/c(x)}$, differentiate with respect to t, and take into account equation (5) and the fact that $[f]|_{t=0} = 2f(0)$. Then we obtain

$$2f(0)\, P(x, t) + \int_{-x}^{x} P(x, \tau)\, f'(t-\tau)\, d\tau + \frac{1}{2}[f'(t-x) + f'(t+x)] = 0, \tag{6б}$$

where by $P(x,t)$ we mean $\sqrt{c(0)/c(x)}\; W(x,t)$, and by $f'(t)$ the derivative of f taken at the continuity points of f; $f(0) \equiv f(+0)$. A similar way of deriving the integral equation is presented in [4].

We introduce the function

$$V(x,t) \equiv \int_0^x \frac{W(\xi,t)}{c(\xi)}\,d\xi. \tag{7}$$

Clearly $V(x,t)$ is a classical function, and $V(x,t) \equiv 0$ for $x < |t|$.

Now we apply the operator

$$\frac{\partial}{\partial t}\int_0^x \frac{1}{c(\xi)}(\cdot)\,d\xi.$$

to both sides of equation (6). Then we get

$$F(x,t) = \frac{\partial}{\partial t}\int V(x,s)f(t-s)\,ds$$

$$= 2f(0)V(x,t) + \int_{-x}^{x} V(x,s)f'(t-s)\,ds, \tag{8}$$

where $F(x,t)$ denotes

$$F(x,t) \equiv \int_0^x \frac{u_t(\xi,t)}{c(\xi)}\,d\xi,$$

and $f'(t)$ is the derivative of $f(t)$, taken at its continuity points, i.e. $f'(t)$ is a classical function and $f'(t) = df(t)/dt$ for $t \neq 0$.

It is easy to compute $F(x,t)$ explicitly for $x > |t|$. Indeed, in the indicated region

$$F_x = \frac{u_t(x,t)}{c(x)} = 0$$

by virtue of (3), and

$$F_t = \int_0^x \frac{u_{tt}(\xi,t)}{c(\xi)}\,d\xi = \int_0^x \frac{\partial}{\partial\xi}\left(\frac{u_\xi(\xi,t)}{c(\xi)}\right)d\xi = \frac{1}{c(x)}[u_x(x,t) - u_x(0,t)] \equiv 0.$$

Hence $F(x,t) = \text{const}$.

Let us find the magnitude of this constant, observing to this end that for $x > 0$

$$\lim_{t\to+0} F(x,t) = \lim_{t\to+0}\int_0^x \frac{u_t^+(\xi,t)}{c(\xi)}\,d\xi = \frac{1}{2}\lim_{t\to+0}\int_{-x}^{x}\frac{u_t^+(\xi,t)}{c(\xi)}\,d\xi.$$

In the last equation we used the even continuation of $u^+(s,t)$ in x. Making use of problem (IV), we find

$$\lim_{t\to+0} F(x, t) = -\frac{1}{2}\int_{-x}^{x} 2\frac{c(0)}{c(\xi)}\delta(\xi)\, d\xi = -1,$$

whence

$$F(x, t) \equiv -1 \quad \text{for} \quad x > |t|.$$

This last assertion, together with (8), now leads to the integral equation

$$-2f(0)V(x, t) - \int_{-x}^{x} f'(t-s)V(x, s)\, ds = 1 \quad \text{for} \quad |t| < x. \tag{F}$$

We should note that we refer to (F) as an integral equation because of the weight of tradition; it is really a family of integral equations for $V(x,s)$, depending on the parameter x. Let us introduce the operator D_f^x, defined by

$$D_f^x\varphi(t) \equiv -2f(0)\varphi(t) - \int_{-x}^{x} f'(t-s)\varphi(s)\, ds. \tag{9}$$

The operator D_f^x is defined on the space of square-integrable functions on the interval $(-x,x)$. Thus equation (F) may be written in the compact form

$$D_f^x V(x, t) = 1.$$

7. Let us show that the function $c(x)$ can be recovered in an unambiguous way from $V(x,t)$ (see the arrow D $\to$ G on the diagram). The proof is carried out under the assumption that problem I is solvable.

Let $V(x,t)$ be defined in terms of $c(x)$ as in formula (7). Then, given that $W(x,t)$ is a solution of problem (III), it is easy to show that $V(x,t)$ is a solution of problem V (that is, of equation (10) subject to condition V):

$$V_{tt} = V_{xx} + \frac{c'}{c}V_x = \frac{1}{c}(cV_x)'_x, \tag{10}$$

$$V|_{x=0} = 0; \quad V_x|_{x=0} = \frac{\delta(t)}{c(0)}. \tag{V}$$

Indeed, the condition on V for $x = 0$ is obviously true. Let us show that V is a solution of equation (10). We have

$$V_{tt} = \int_0^x \frac{1}{c(\xi)} W_{tt}\, d\xi = \int_0^x \left(\frac{W_\xi}{c}\right)'_\xi d\xi = \frac{W'_x}{c(x)}$$

$$= \frac{1}{c}\left\{c(x)\left[\int_0^x \frac{W(\xi, t)}{c(\xi)}\, d\xi\right]'_x\right\}'_x = \frac{1}{c}[cV'_x]'_x.$$

Consider the integral

$$R(x) = \int_{-\infty}^{\infty} V(x, t)\, dt. \tag{11}$$

Clearly $R(x)$ satisfies the equation

$$\frac{1}{c}[cR']' = 0,$$

whence $cR' = \text{const}$. From (11) and the condition on $V_x|_{x=0}$ we find that $cR' = 1$. Hence

$$c(x) = \frac{1}{R'(x)}. \tag{12}$$

There is another possible way of determining $c(x)$ from $V(x,t)$. Observe that it is not too difficult to determine the jump of the function $V(x,t)$ across the characteristic $x = t$; this can be calculated from problem (V) by means of the theory of characteristics. One finds that

$$V(x, x) = \frac{1}{2\sqrt{c(x)\,c(0)}},$$

whence

$$c(0) = \frac{1}{2V(0, 0)},$$

$$c(x) = \frac{V(0, 0)}{2V(x, x)^2}. \tag{13}$$

In the above formulas, $V(x,x)$ denotes the limit of $V(x,t)$ as $t \to x - 0$.

8. Let us show that the solvability of the inverse problem (I) implies the validity of the following condition.

The inequality $(D_f^x \varphi, \varphi) > 0$ *is valid for all*

$x > 0$ *and* $\varphi \in L_{2(-x,x)}$, *as long as* $\varphi \neq 0$. (K)

In other words, the positive definiteness of the operator D_f^x is necessary for the solvability of problem I. We must study some properties of the operator D_f^x. Its definition and the fact that $f'(t)$ is even imply that D_f^x may be represented in the form

$$D_f^x = -2f(0)\,E^x + K^x,$$

where E^x is the identity, and K^x is an integral operator with symmetric kernel. Since under the assumptions of the problem we have $f(0) < 0$, and since K^x is small in norm for small x, it is clear that D_f^x will be positive definite for small enough x, and that all eigenvalues of D_f^x, in particular the least one, $\lambda_0(x)$, are positive. Suppose that for some x_1 the operator D_f^x is not positive definite. This means, in particular, that its least eigenvalue $\lambda_0(x_1) \leqslant 0$. But now the operator D_f^x, and so its least eigenvalue, depend continuously on x. Thus there exists a point x_0 for which $\lambda_0(x_0) = 0$. Then there is a function $\varphi_0 \neq 0$ such that $D_f^{x_0}\varphi_0 = 0$, or

$$\frac{d}{dt}\int_{-x_0}^{x_0} f(t-s)\,\varphi_0(s)\,ds = 0. \tag{14}$$

9. Let us look at the function

$$\Phi_t(x,\ t) \equiv \int_{-x_0}^{x_0} u_t(x,\ t-s)\,\varphi_0(s)\,ds, \tag{15}$$

which may be written in the form of a convolution

$$\Phi_t(x,\ t) = u_t(x,\ t) * \varphi_0(t),$$

where $\varphi_0(t)$ denotes the eigenfunction φ_0, continued to be identically null outside the interval $(-x_0, x_0)$. The differentiation in (15) is understood to be in the sense of distributions; for instance, in a neighborhood of $t = x$ we have

$$u_t(x,\ t) = -\sqrt{\frac{c(x)}{c(0)}}\,\delta(t-x) + \ldots, \tag{16}$$

where the dots denote a classical function; the magnitude of the jump of the function $u(x,t)$ for $t = x$, which is taken into account in (16), is easily computed by means of the theory of characteristics.

Clearly $\Phi_t(x,t)$ satisfies equation (1) and the conditions

$$\Phi_t|_{x=0} = 0, \quad \Phi_{tx}|_{x=0} = 0 \quad \text{for } t \in (-x_0,\ x_0).$$

Therefore, by virtue of the hyperbolicity of equation (1),

$$\Phi_t(x,\ t) \equiv 0 \quad \text{for } x - x_0 < t < x_0 - x, \quad x < x_0.$$

10. The eigenfunction $\varphi_0(t)$ may be taken to be either even or odd, since it follows from the evenness of the function $f'(t)$ that the operator D_f^x commutes with the operator S^x, where $S^x\varphi(t) \equiv \varphi(-t)$. Suppose we take $\varphi_0(t)$ to be even. Then

$$\Phi_t(x,\ 0) = -\int_{-x_0}^{x_0} u_s(x,\ -s)\,\varphi_0(s)\,ds = -2\int_0^{x_0} u_s(x,\ s)\,\varphi_0(s)\,ds \equiv 0$$

for $x < x_0$. Or, remembering (16) and the fact that $u|_{|s|<x} \equiv 0$,

$$-\sqrt{\frac{c(x)}{c(0)}}\,\varphi_0(x) + \int_{x+0}^{x_0} u_s(x,\ s)\,\varphi_0(s)\,ds = 0.$$

Now this last equation is a homogeneous Volterra integral equation of the second kind for the function φ_0, whence $\varphi_0 \equiv 0$. Suppose next that the function φ_0 is odd. Then we consider the function

$$\Phi(x, t) \equiv \int_{-x_0}^{x_0} u(x, t-s)\,\varphi_0(s)\,ds.$$

From the preceding it follows that $\Phi_t(x,t) \equiv 0$ for $x - x_0 < t < x_0 - x$, $x < x_0$, so that $\Phi(x,t) = \Phi(x)$. Using equation (1), we find that in the above-mentioned region

$$\frac{\Phi'(x)}{c(x)} = \text{const}.$$

Because $\Phi'(0) = \int_{-x_0}^{x_0} u_x(x, t-s)|_{x=0}\varphi_0(s)\,ds$, we have $\Phi'(x) \equiv 0$ and $\Phi(x) = \text{const}$. Now observe that $\Phi(x_0) = \int_{-x_0}^{x_0} u(x_0, -s)\varphi_0(s)\,ds = 0$, because the function $u(x_0, t)$ is a classical one, and $u(x,t) \equiv 0$ for $|t| < x$. Therefore $\Phi(x) \equiv 0$. This means that

$$\int_{-x_0}^{x_0} u(x, -s)\,\varphi_0(s)\,ds = -2\int_{x}^{x_0} u(x, s)\,\varphi_0(s)\,ds \equiv 0$$

and $\varphi_0 \equiv 0$, being a solution of a homogeneous Volterra integral equation of the first kind (in the derivation of the last assertion we used the fact that $u(x, x+0) \neq 0$).

In summary, the violation of condition (K) entails the existence of an eigenfunction $\varphi_0(t)$ of the operator $D_f^{x_0}$, and the solvability of the inverse problem (I) entails that $\varphi_0(t)$ must be zero, which is absurd. This proves the assertion A $\to$ C.

11. According to the theory of integral equations with symmetric kernel, condition (K) implies the unique solvability of equation (F) (condition (K) means that we are staying away from the spectrum in equation (F)). This proves the implications C $\to$ D and C $\to$ E.

12. Let us begin to prove that condition (K) entails the possibility of constructing a solution $c(x)$ of problem (I) by means of a function $V(x,t)$ satisfying equation (F). The function $V(x,t)$, being a solution of equation (F), is constructed only for $|t| < x$. Let us define $V(x,t)$ for $|t| > x$ by setting it equal to zero. Thus we construct a function $V(x,t)$ which is classical and piecewise continuous for all t and $x > 0$. It is easily verified that $V(x,t)$ is an even function of t. Let us show first that $V(x,x) \neq 0$ for all x. Indeed, suppose that $V(x_0, x_0) = 0$. Then

$$D_f^{x_0}V(x_0, t) = -\frac{\partial}{\partial t}\int f(t-s)\,V(x_0, s)\,ds = \int f(t-s)\frac{\partial}{\partial s}V(x_0, s)\,ds = 1.$$

But $V(x_0, s)$ is continuous for all s, whence $\partial V(x_0, s)/\partial s$ is a classical function for all s, and

$$\frac{\partial}{\partial t} D_f^{x_0} V(x_0, t) = -\frac{\partial}{\partial t} \int_{-x_0}^{x_0} f(t-s) \frac{\partial}{\partial s} V(x_0, s)\, ds = D_f^{x_0} V_t(x_0, t) = 0.$$

Since we are staying away from the spectrum of the operator $D_f^{x_0}$ (this follows from condition (K)), we find that $\partial V(x_0, t)/\partial t \equiv 0$, and since $V(x_0, x_0) = 0$ it follows that $V(x_0, t) = 0$, an impossibility because by construction $V(x_0, t)$ is supposed to be a solution of the inhomogeneous equation (F).

13. In §7, we proved that the solvability of problem (I) implies that $V(x, t)$ must be a solution of problem V; $c(x)$ is then found from formula (13). Here we show that the same conclusion can be obtained as a consequence of condition (K). Let $V(x, t)$ be the function constructed in §12. Define $c(x)$ in terms of V according to (13):

$$c(x) = \frac{V(0, 0)}{2V(x, x)^2}.$$

Let us define

$$\omega(x, t) \equiv \frac{\partial^2}{\partial t^2} V(x, t) - \frac{1}{c(x)} \frac{\partial}{\partial x} [c(x) V_x(x, t)].$$

It is easily verified that $\omega(x, t)$ is a classical function. Indeed, because of the smoothness of the kernel $f'(t-s)$, the discontinuities of the function V can only be located at the points $x = \pm t$, and since $V(x, t)$ is even, our attention may be restricted to a neighborhood of $x = t$. Let

$$V(x, t) = V(x, x)\, \varepsilon(x-t) + \dots, \quad \text{where} \quad \varepsilon(t) = \begin{cases} 1, & t > 0, \\ 0, & t < 0; \end{cases}$$

dots denote less singular terms. Substituting this representation into the expression for ω, we find

$$\begin{aligned}
\omega(x, t) &= \delta'(x-t) V(x, x) - \frac{1}{c(x)} \frac{\partial}{\partial x} \left\{ c(x) \frac{\partial}{\partial x} [V(x, x)\, \varepsilon(x-t)] \right\} + \dots \\
&= \delta(x-t) V(x, x) - V(x, x) \delta'(x-t) - \delta(x-t) \frac{1}{c(x)} \\
&\quad \times \frac{\partial}{\partial x} [c(x) V(x, x)] \delta(x-t) - \left[\frac{\partial}{\partial x} V(x, x) \right] \delta(x-t) + \dots \\
&= -\left\{ \frac{1}{c(x)} \frac{\partial}{\partial x} [c(x) V(x, x)] + \frac{\partial}{\partial x} V(x, x) \right\} \delta(x-t) + \dots \\
&= -\frac{1}{c(x) V(x, x)} \frac{\partial}{\partial x} [c(x) V^2(x, x)]\, x\delta(x-t) + \dots
\end{aligned}$$

It is clear from (13) that the coefficient of $\delta(x-t)$ is zero. In a similar fashion, one can show that the subsequent terms of the expansion of $V(x, t)$ near the point $x = t$

also contribute classical terms only. Let us prove that $\omega(x,t) \equiv 0$ for all t, $x > 0$. For $|t| > x$ this is obvious. In order to establish the result for $|t| < x$, we write equation (F):

$$D_f^x V(x, t) = -\frac{\partial f(t)}{\partial t} * V(x, t) = 1$$

for $|t| < x$, and apply the operator

$$\frac{\partial^2}{\partial t^2} - \frac{1}{c(x)} \frac{\partial}{\partial x} c(x) \frac{\partial}{\partial x}.$$

to both sides of the equation. The result is

$$-\frac{\partial f(t)}{\partial t} * \omega(x, t) = D_f^x \omega(x, t) = 0,$$

whence $\omega(x,t) = 0$ for $|t| < x$. Furthermore, it is clear that

$$V(x, t)|_{x=0} = 0,$$

and

$$\left.\frac{\partial V}{\partial x}\right|_{x=0} = V(x, x)[\delta(x-t) + \delta(x+t)]|_{x=0} = 2V(0, 0)\delta(t) = \frac{1}{c(0)}\delta(t).$$

This proves that $V(x,t)$ is a solution of problem V.

14. Let us employ the function V to construct the function $u^+(x,t)$. To this end we first define

$$W(x, t) \equiv c(x) V'_x(x, t).$$

One may easily verify that $W(x,t)$ is a solution of problem (III). Next we construct $u(x,t)$ according to formula (6):

$$u(x, t) \equiv \int W(x, s) f(t-s)\, ds.$$

Obviously $u(x,t)$ is a solution of problem (II). Let us show that $u(x,t) \equiv 0$ for $x > |t|$. Consider the function

$$F(x, t) = \frac{\partial}{\partial t} \int_0^x \frac{u(\xi, t)}{c(\xi)}\, d\xi =$$

$$= \frac{\partial}{\partial t} \int ds \left[\int_0^x \frac{W(\xi, s)}{c(\xi)}\, d\xi \right] \cdot f(t-s) = \frac{\partial}{\partial t} \int ds f(t-s) V(x, s).$$

Because of equation (F), $F(x,t) \equiv -1$ for $|t| < x$. Differentiating $F(x,t)$ with respect to x, we find that $u_t(x,t) \equiv 0$ for $|t| < x$. This means that $u(x,t) \equiv u(x)$ for $|t| < x$. But by construction $u(x,t)$ is odd, which means that it must equal zero in the region $|t| < x$.

Finally, we construct the function

$$u^{+}(x, t)=\begin{cases} u(x, t) & \text{for} \quad t \geqslant 0, \\ 0 & \text{for} \quad t \leqslant 0. \end{cases}$$

Clearly the pair of functions $c(x)$, $u^{+}(x,t)$ satisfies all conditions of problem (I). This proves the solvability of problem (I), given the validity of condition (K).

Bibliography

1. M.G. Kreĭn, *On a method of effective solution of an inverse problem*, Dokl. Akad. Nauk SSSR **94** (1954), 987–990. (Russian) MR **16**, 38.
2. I.M. Gel'fand and B.M. Levitan, *On the determination of a differential equation from its spectral function*, Izv. Akad. Nauk SSSR Ser. Mat. **15** (1951), 309–360; English transl., Amer. Math. Soc. Transl. (2) **1** (1955), 253–304. MR **13**, 558; **17**, 489.
3. L.D. Faddeev, *The inverse problem in the quantum theory of scattering*, Uspehi Mat. Nauk **14** (1959), no. 4 (88), 57–119; English transl., J. Mathematical Phys. **4** (1963), 72–104. MR **22** #1344;**26** #7328.
4. B.S. Pariĭskiĭ, *The inverse problem for the wave equation, with influence on intensity*, Computational Seismology, no. 2, "Nauka", Moscow, 1969. (Russian)
5. A.S. Blagoveščenskiĭ, *On an inverse problem in the theory of seismic wave propagation*, Problemy Mat. Fiz., no. 1, Izdat. Leningrad. Univ., Leningrad, 1966, (Russian) pp. 68–81; English transl., Topics in Math. Phys., no. 1, Consultants Bureau, New York, 1967, 55–67.

Translated by:
H. Flaschka

AN INVERSE BOUNDARY PROBLEM IN THE THEORY OF PROPAGATION OF WAVES IN AN ANISOTROPIC MEDIUM

A. S. BLAGOVEŠČENSKIĬ

UDC 517 944/947

Abstract. A nonstationary inverse problem is considered concerning determination of the coefficients of the equation $u_{tt} = Lu$, where L is a homogeneous elliptic operator of the form

$$L = c^2 \frac{\partial^2}{\partial z^2} + 2b \frac{\partial^2}{\partial z \partial x} + a \frac{\partial^2}{\partial x^2} \tag{1}$$

or

$$L \equiv \frac{\partial}{\partial z}\left(c^2 \frac{\partial}{\partial z}\right) + \frac{\partial}{\partial z}\left(b \frac{\partial}{\partial x}\right) + \frac{\partial}{\partial x}\left(b \frac{\partial}{\partial z} + a \frac{\partial}{\partial x}\right). \tag{2}$$

It is shown that in case (1) the coefficients c, b, and a can be found if the integrals

$$F_k(t) \equiv \frac{1}{k!} \int_{-\infty}^{\infty} f(x, t)\, x^k dx, \quad (k = 0, 1, 2) \tag{3}$$

are known. It is a great surprise that in case (2) the problem has a completely different character and prescription of the moments (3) does not determine the unknown coefficients. In the case of an operator of the form (2) the problem is also investigated in detail.

Bibliography: 6 items.

Two problems concerning determination of the coefficients of a hyperbolic equation describing the process of wave propagation in an anisotropic medium are considered in this paper. It is curious that in spite of apparent similarity in the statement of these two problems, qualitatively different results are obtained for them. The methods of analyzing them are also essentially different, coinciding only at the first step (conversion to one-dimensional problems by introducing moments into consideration).

Let us formulate these problems. Suppose it is known that in the halfplane $z > 0$ the function $u(x, z, t)$ is a solution of the differential equation

$$u_{tt} = Lu \tag{1}$$

which satisfies the conditions

$$u|_{t<0} \equiv 0, \tag{2}$$

$$u_z|_{z=0} = \delta(x, t). \tag{3}$$

Let $f(x, t)$ be the value of the function u for $z = 0$:

$$u|_{z=0} \equiv f(x, t). \tag{4}$$

AMS 1970 *subject classification*. Primary 35L20, 35R30; Secondary 44A35, 44A50, 45D05.

L is an elliptic operator of one of the following kinds: in the first problem (which we shall call nonselfadjoint)

$$Lu \equiv c^2 u_{zz} + 2\tilde{b} u_{zx} + a u_{xx}; \tag{5}$$

in the second problem (selfadjoint) the operator L is assumed to be formally selfadjoint:

$$Lu \equiv \frac{\partial}{\partial z}(c^2 u_z) + \frac{\partial}{\partial z}(\tilde{b} u_x) + \frac{\partial}{\partial x}(\tilde{b} u_z) + \frac{\partial}{\partial x}(a u_x). \tag{6}$$

In both problems a, $\tilde{b}$ and c are unknown smooth functions of the variable z. Some information about the function $f(x, t)$ is assumed to be known.

We prove that the nonselfadjoint problem is uniquely solvable if the first three moments $F_k(t)$ ($k = 0, 1, 2$) relative to x of the function $f(x, t)$

$$F_k(t) \equiv \frac{1}{k!} \int_{-\infty}^{\infty} f(x, t)\, x^k dx. \tag{7}$$

are prescribed. The results are very surprising concerning the selfadjoint problem, for which it is established that the functions $a(z)$ and $\tilde{b}(z)$ cannot be found uniquely even if $f(x, t)$ is completely known. All of the information about the coefficients of operator (6) which can be obtained in the process of solving the problem is exhausted by the following quantities: 1) the function $c(z)$, 2) the determinant $D(z) \equiv a(z)c^2(z) - \tilde{b}^2(z)$, and 3) the value of $\tilde{b}(0)$. Moreover, prescribing the first three moments F_0, F_1 and F_2 is sufficient to determine the aforementioned quantities, it being necessary for solvability of the problem to impose very strong restrictions on F_1 (F_1 is found from F_0 to within a constant factor equal to $\tilde{b}(0)$). The fact that such an essentially nonlinear quantity as the determinant $D(z)$ satisfies a linear integral equation proved to be unexpected.

We shall begin by considering the selfadjoint problem. The first section is devoted to several auxiliary constructions (possibly of independent interest). The results formulated above for the selfadjoint problem are proved in the second and third sections. In the fourth section we develop the analysis of the nonselfadjoint problem very briefly, since it is conceptually very similar to considerations developed in [1], [2] and particularly [3].

§1. The Wronskian convolution determinant

Let $u_i(z, t)$, $i = 1, 2$, be two solutions of the equations

$$p(z) u_{itt} + q(z) u_{it} + r(z) u_i = \frac{\partial}{\partial z}(c^2 u_{iz}) + f_i(z, t) \tag{8}$$

which are generalized functions of t in K'_+, i.e. functions of K'[1] whose support is concentrated on the semi-axis $t \geqslant 0$. We regard the variable z as a parameter; the functions u_i are assumed to be continuously differentiable twice with respect to z

[1] K' is the space conjugate to the space K of infinitely differentiable functions of compact support.

(of course in the sense of generalized function theory); the $f_i(z,t)$ are generalized functions of K'_+ which are continuous with respect to z. The coefficients p, q and r are continuous functions only of z; $c(z)$ is a continuously differentiable, everywhere positive function of z.

As is well known [4], for any pair of functions of K'_+, their convolution is defined (the convolution is always implicit with respect to t). Let us compose formally the determinant of the matrix

$$\begin{pmatrix} u_1 & u_2 \\ u_{1z} & u_{2z} \end{pmatrix},$$

and then substitute the convolution for the products of the matrix elements which are obtained when composing the determinant. We call the resulting generalized function, which depends on the parameter z, a Wronskian convolution determinant, or concisely, a convolution Wronskian, and denote it by the symbols

$$W^* = [u_1(z) * u_2(z)] = \begin{vmatrix} u_1 & u_2 \\ u_{1z} & u_{2z} \end{vmatrix}^* \equiv u_1 * u_{2z} - u_2 * u_{1z} \tag{9}$$

(we omit the dependence of the functions $u_1(z)$ and $u_2(z)$ on t from the notation, tacitly assuming it). In what follows the functions u_k will, as a matter of fact, often depend on several parameters $z, z_1, z_2, \ldots, z_n$, the first of the functions depending on not more than two arguments: $u_1 = u_1(z, z_1)$. The notation

$$[u_1(z, z_1) * u_2(z_1, z_2, \ldots, z_n)]$$

means that when forming the convolution Wronskian, u_1 and u_2 are differentiated with respect to the second argument z_1 of the function u_1.

Let us return to the situation described at the beginning of the section and make an evaluation of the Wronskian determinant

$$W^*(z) = [u_1(z) * u_2(z)].$$

For this purpose we introduce a differential equation which the function $W^*(z)$ satisfies, for which we multiply W^* by $c^2(z)$ and differentiate with respect to z. Utilizing the rule for differentiating a convolution with respect to a parameter, we obtain the usual rule for differentiating a determinant. Therefore

$$\frac{\partial}{\partial z}(c^2(z) W^*) = \frac{\partial}{\partial z}\begin{vmatrix} u_1 & u_2 \\ c^2 u_{1z} & c^2 u_{2z} \end{vmatrix}^* = \begin{vmatrix} u_{1z} & u_{2z} \\ c^2 u_{1z} & c^2 u_{2z} \end{vmatrix}^* + \begin{vmatrix} u_1 & u_2 \\ \frac{\partial}{\partial z}(c^2 u_{1z}) & \frac{\partial}{\partial z}(c^2 u_{2z}) \end{vmatrix}^*.$$

The first addend obviously equals zero; in order to evaluate the second addend, let us use differential equations (8) for the functions u_i. We obtain

$$\frac{\partial}{\partial z}(c^2(z)\,W^*)=\begin{vmatrix} u_1 & u_2 \\ pu_{1tt} & pu_{2tt}\end{vmatrix}^*+\begin{vmatrix} u_1 & u_2 \\ qu_{1t} & qu_{2t}\end{vmatrix}^*+\begin{vmatrix} u_1 & u_2 \\ ru_1 & ru_2\end{vmatrix}^*-\begin{vmatrix} u_1 & u_2 \\ f_1 & f_2\end{vmatrix}^*.$$

Let us consider the first addend:

$$\begin{vmatrix} u_1 & u_2 \\ pu_{1tt} & pu_{2tt}\end{vmatrix}^*=p\,(u_1*u_{2tt}-u_2*u_{1tt})=p\,\frac{\partial^2}{\partial t^2}(u_1*u_2-u_2*u_1)=0.$$

The two subsequent addends analogously equal zero; therefore

$$\frac{\partial}{\partial z}(c^2W^*)=-\begin{vmatrix} u_1 & u_2 \\ f_1 & f_2\end{vmatrix}^*=f_1*u_2-f_2*u_1, \tag{10}$$

and

$$W^*=\frac{1}{c^2(z)}h+\frac{1}{c^2(z)}\int(f_1*u_2-f_2*u_1)\,dz, \tag{11}$$

where h is an arbitrary generalized function $h\in K'_+$ which does not depend on z.

In what follows we shall be interested in the value of the convolution Wronskian in two special cases. Let $p\equiv 1$ and $q=r\equiv 0$. Equation (8) is considered on the semi-axis $z>0$. Let the boundary condition

$$\left.\frac{\partial u_i}{\partial z}\right|_{z=0}=0 \tag{12}$$

be imposed on $u_i(z)$. We call the function satisfying the conditions

$$\left.\begin{aligned} &G_{tt}=\frac{\partial}{\partial z}(c^2(z)\,G_z)\quad\text{for}\quad t>0,\quad z>0,\\ &G_z|_{z=0}=0,\\ &G|_{t=0}=0,\\ &G_t|_{t=0}=\delta(z-z_1)\end{aligned}\right\} \tag{13}$$

the Green's function $G(z,z_1)$. The following properties of the Green's function are very well known:

1) Symmetry:

$$G(z,\ z_1)=G(z_1,\ z). \tag{14}$$

2) If $G(z,z_1)$, defined above for $t\geqslant 0$, is continued to negative t as identically zero, $G(z,z_1)$ becomes a solution of the problem

$$\left.\begin{aligned} &\frac{\partial^2}{\partial t^2} G = \frac{\partial}{\partial z}(c^2 G_z) + \delta(z - z_1)\delta(t), \\ &G_z|_{z=0} = 0, \\ &G|_{t<0} \equiv 0, \qquad \text{or} \quad G \in K'_+. \end{aligned}\right\} \tag{15}$$

Let us evaluate the convolution Wronskian $[G(z, z_1) * G(z_1, z_2)]$ for $z, z_1, z_2 > 0$. Applying formulas (11), (14) and (15), we find

$$[G(z, z_1) * G(z_1, z_2)] = \frac{h(z, z_2)}{c^2(z_1)} + \frac{1}{c^2(z_1)} \int dz_1 \{\delta(z - z_1)\delta(t) * G(z_1, z_2)$$
$$- \delta(z_1 - z_2)\delta(t) * G(z, z_1)\} = \frac{h(z, z_2)}{c^2(z_1)} + \frac{1}{c^2(z_1)}[\varepsilon(z_1 - z) - \varepsilon(z_1 - z_2)]\, G(z, z_2),$$

where $\varepsilon(x)$ is the Heaviside function

$$\varepsilon(x) = \begin{cases} 1 & \text{for } x \geqslant 0, \\ 0 & \text{for } x < 0. \end{cases}$$

Let us use the boundary condition

$$G_{z_1}(z, z_1)|_{z_1=0} = G_{z_1}(z_1, z_2)|_{z_1=0} = 0$$

to evaluate the function $h(z, z_2)$. On the one hand,

$$[G(z, z_1) * G(z_1, z_2)]|_{z_1=0} = \left|\begin{matrix} G(z, z_1), & G(z_1, z_2) \\ G_{z_1}(z, z_1), & G_{z_1}(z_1, z_2) \end{matrix}\right|^{*} \Bigg|_{z_1=0} = 0.$$

On the other hand,

$$[G(z, z_1) * G(z_1, z_2)]|_{z_1=0} = \frac{h(z, z_2)}{c^2(0)}$$
$$+ \frac{1}{c^2(0)}[\varepsilon(-z) - \varepsilon(-z_2)]\, G(z, z_2) = \frac{h(z, z_2)}{c^2(0)},$$

from which we obtain $h(z, z_2) \equiv 0$ and

$$[G(z, z_1) * G(z_1, z_2)] = \frac{1}{c^2(z_1)}[\varepsilon(z_1 - z) - \varepsilon(z_1 - z_2)] \cdot G(z, z_2). \tag{16}$$

From (16) we easily obtain

$$[G(z, z_1) * G(z_1, 0)] = -\frac{1}{c^2(z_1)}\varepsilon(z - z_1)\, G(z, 0), \tag{17}$$

$$[G(0, z_1) * G(z_1, z_2)] = \frac{1}{c^2(z_1)}\varepsilon(z_2 - z_1)\, G(0, z_2) \tag{18}$$

and

$$[G(0, z_1) * G(z_1, 0)] \equiv 0. \tag{19}$$

However, formula (19) can also be easily obtained from the symmetry of Green's function. By $G(z,0) = G(0,z)$ we everywhere understand the limit $G(z,+0)$.

Let us introduce one more useful formula. Let $\lambda = \lambda(z)$ be some function of z (and only of z). It is then obvious that

$$[u_1 * \lambda u_2] = \begin{vmatrix} u_1 & \lambda u_2 \\ u_{1_z} & (\lambda u_2)_z \end{vmatrix}^* = \begin{vmatrix} u_1 & \lambda u_2 \\ u_{1_z} & \lambda u_{2z} \end{vmatrix}^* + \begin{vmatrix} u_1 & 0 \\ u_{1_z} & \lambda_z u_2 \end{vmatrix}^* = \lambda [u_1 * u_2] + \lambda_z u_1 * u_2.$$

In particular,

$$[G(z, z_1) * \varepsilon(z_1 - z_3) G(z_1, z_2)] = \varepsilon(z_1 - z_3)[G(z, z_1) * G(z_1, z_2)] + \delta(z_1 - z_3) \dot{G}(z, z_1) * G(z_1, z_2)$$

or, utilizing (16),

$$[G(z, z_1) * \varepsilon(z_1 - z_3) G(z_1, z_2)] = \frac{1}{c^2(z_1)} \varepsilon(z_1 - z_3)[\varepsilon(z_1 - z) - \varepsilon(z_1 - z_2)] G(z, z_2) + \delta(z_1 - z_3) G(z, z_1) * G(z_1, z_2). \tag{20}$$

REMARKS. 1. Strictly speaking, formula (11) does not quite imply the validity of equalities (16)–(20), since when deriving formula (11) it was assumed that the functions $f_1(z)$ and $f_2(z)$ are continuous as functions of the argument z. This does not hold for the functions occurring in (16)–(20). Nevertheless, formulas (16)–(20) can be proved rigorously. We shall not develop the proof here in order not to overload the paper with nonessential details.

2. The results concerning the convolution Wronskian can of course be easily generalized to the cases of equations of higher order and systems.

3. Formula (11) can also be easily proved if one observes that the convolution Wronskian $W^*(z)$ is the Fourier preimage of the ordinary Wronskian determinant for the functions $\tilde{u}_1(z,s)$ and $\tilde{u}_2(z,s)$, where $\tilde{u}_k(z,s)$ is the Fourier transform of the function $u_k(z,t)$ with respect to t.

§2. Selfadjoint inverse problem. I.

Let us consider the problem of finding the coefficients a, $\tilde{b}$ and c^2 in the equation

$$\frac{\partial^2 u}{\partial t^2} = \frac{\partial}{\partial z}(c^2 u_z) + \frac{\partial}{\partial z}(\tilde{b} u_x) + \tilde{b} u_{zx} + a u_{xx} \quad (z > 0)$$

if it is known that

$$\left.\begin{aligned} u_z|_{z=0} = 0, \ u|_{t=0} = 0, \ u_t|_{t=0} = \delta(z)\delta(x), \\ u|_{z=0} = f(x, t), \end{aligned}\right\} \tag{21}$$

where $f(x,t)$ is a prescribed function; by $\delta(z)$ we understand $\lim_{z_1 \to +0} \delta(z - z_1)$; we assume a, $\tilde{b}$ and c to be sufficiently smooth functions of z; the differential operator on the right side of (6) is assumed elliptic.

Let us introduce the moments of the function $u(x, z, t)$:

$$V_k(z, t) \equiv \frac{1}{k!} \int_{-\infty}^{\infty} u(x, z, t) x^k dx. \tag{22}$$

It is easy to show that the $V_k(z, t)$ are solutions of the following problems:

$$\begin{gathered} V_{ktt} = \frac{\partial}{\partial z}(c^2 V_{kz}) - \frac{\partial}{\partial z}(\tilde{b} V_{k-1}) - \tilde{b} V_{k-1z} + a V_{k-2}, \\ V_{kz}|_{z=0} = 0, \ V_k|_{t=0} = 0, \\ V_{kt}|_{t=0} = \delta(z)\delta_{k0}, \ V_k|_{z=0} = F_k(t), \end{gathered} \tag{23}$$

where δ_{k0} is the Kronecker symbol and[2)]

$$F_k(t) \equiv \frac{1}{k!} \int_{-\infty}^{\infty} f(x, t) x^k dx.$$

The formulation of problem (23) makes sense both for $k = 0$ and for $k = 1$ if we agree to assume that $V_{-2} \equiv V_{-1} \equiv 0$.

Let $k = 0$. Then we obtain the problem

$$\left.\begin{gathered} V_{0tt} = \frac{\partial}{\partial z}(c^2(z) V_{0z}), \\ V_0|_{t=0} = 0, \ V_{0t}|_{t=0} = \delta(z), \ V_{0z}|_{z=0} = 0, \\ V_0|_{z=0} = F_0(t). \end{gathered}\right\} \tag{I}$$

After the introduction of a new coordinate $y = \int_0^z (1/c(\zeta))d\zeta$ in place of z and a new unknown function $c_1(y) = 1/c(y)$, this last problem coincides exactly with problem I of article [5] in the present collection. Let us suppose that the function $F_0(t)$ satisfies condition (K) of article [5], i.e., the necessary and sufficient condition for the solvability of problem (I), which was obtained first by M.G. Kreĭn [6]. Problem (I) has been considered in sufficient detail, in the papers cited here, for example. We shall assume that problem (I) has already been solved and by the same token the functions $c(z)$ and $V_0(z, t)$ found. Let us further assume that we know the Green's function $G(z, z_1)$, which was defined in §1 as the solution of the direct problem (13), into whose formulation the function $c(z)$, already found, enters.

It is obvious that the formula[3]

$$V_0(z) = G(z, 0) = G(0, z) \tag{24}$$

[2)] Let us note that the functions $u(x, z, t)$, $f(x, t)$ and $F_k(t)$ introduced in this section differ by the incidental factor $1/c^2(0)$ from the functions denoted by the same symbols in the Introduction, as is easily shown by comparing the formulations of problems (6), (2)–(4) and (6), (21).

[3)] Just as in §1, the dependence of the function on t is not written out explicitly below but is implicit.

is valid. Our basic aim is to extract the most possible information concerning the functions $\tilde{b}(z)$ and $a(z)$ from considering the successive moments $V_1(z)$, $V_2(z), \ldots$. The principal role in the considerations which follow will be played by formula (28) (see below), whose derivation we now take up.

Let the function $W(z)$ be a solution of the problem

$$\left.\begin{aligned} W_{tt} &= \frac{\partial}{\partial z}(c^2 W_z) + H(z) \quad (z > 0), \\ W_z|_{z=0} &= 0, \quad W|_{t<0} = 0. \end{aligned}\right\} \tag{25}$$

Then $W(z)$ is found according to the formula

$$W(z) = \int_0^\infty G(z, z_1) * H(z_1)\, dz_1. \tag{26}$$

Let us apply formula (26) to finding the solutions $V_k(z)$ of problems (23) ($k = 1, 2, \ldots$). Then we obtain

$$\begin{aligned} V_k(z) = &-\int_0^\infty G(z, z_1) * \frac{\partial}{\partial z_1}(\tilde{b}(z_1) V_{k-1}(z_1))\, dz_1 \\ &-\int_0^\infty G(z, z_1) * \tilde{b}(z_1) \frac{\partial}{\partial z_1} V_{k-1}(z_1)\, dz_1 + \int_0^\infty G(z, z_1) * V_{k-2}(z_1)\, a(z_1)\, dz_1. \end{aligned} \tag{27}$$

Let us integrate the first addend in (27) by parts, taking into account that the functions $V_{k-1}(z_1)$ and $G(z, z_1)$ tend to zero as $z_1 \to \infty$ by virtue of the hyperbolicity of the equations:

$$\begin{aligned} V_k(z) = {} & \tilde{b}(0) G(z, 0) * V_{k-1}(0) + \int_0^\infty G_{z_1}(z, z_1) * V_{k-1}(z_1) \tilde{b}(z_1)\, dz_1 \\ & - \int_0^\infty G(z, z_1) * V_{k-1\, z_1}(z_1) \tilde{b}(z_1)\, dz_1 + \int_0^\infty G(z, z_1) * V_{k-2}(z_1)\, a(z_1)\, dz_1 \\ = {} & \tilde{b}(0) G(z, 0) * F_{k-1} - \int_0^\infty \{G(z, z_1) * V_{k-1\, z_1}(z_1) - G_{z_1}(z, z_1) * V_{k-1}(z_1)\} \\ & \times \tilde{b}(z_1)\, dz_1 + \int_0^\infty G(z, z_1) * V_{k-2}(z_1)\, a(z_1)\, dz_1 = \tilde{b}(0) G(z, 0) * F_{k-1} \\ & - \int_0^\infty [G(z, z_1) * V_{k-1}(z_1)]\tilde{b}(z_1)\, dz_1 + \int_0^\infty G(z, z_1) * V_{k-2}(z_1)\, a(z_1)\, dz_1. \end{aligned}$$

We have here used the concept of convolution Wronskian which was introduced in §1, and also the boundary condition $V_{k-1}(0) = F_{k-1}$ of problems (23).

Thus we have derived the formula

$$V_k(z) = \tilde{b}(0)\, G(z, 0) * F_{k-1} - \int_0^\infty [G(z, z_1) * V_{k-1}(z_1)]\, \tilde{b}(z_1)\, dz_1 + \int_0^\infty G(z, z_1) * V_{k-2}(z_1)\, a(z_1)\, dz_1, \tag{28}$$

on which all subsequent considerations of the selfadjoint problem are based. Henceforth when writing out formula (28) and its numerous variants, we shall not put down the limits of integration each time, it being implicit that they are always one and the same: from zero to infinity. In what follows the notation $\beta(z) \equiv \tilde{b}(z)/c^2(z)$ will also be useful to us.

Let $k = 1$. In that case formula (28) is valid if it is assumed that $V_{-1}(z) \equiv 0$. Consequently we have

$$\begin{aligned} V_1(z) &= \tilde{b}(0)\, G(z, 0) * F_0 - \int [G(z, z_1) * V_0(z_1)]\, \tilde{b}(z_1)\, dz_1 \\ &= \tilde{b}(0)\, G(z, 0) * F_0 - \int [G(z, z_1) * G(z_1, 0)]\, \tilde{b}(z_1)\, dz_1 \\ &= \tilde{b}(0)\, G(z, 0) * F_0 + \int G(z, 0)\, \varepsilon(z - z_1)\, \beta(z_1)\, dz_1. \end{aligned} \tag{29}$$

Representation (17) for the convolution Wronskian was utilized to derive (29). After setting $z = 0$ in (29) and observing that $V_1(0) = F_1$ and $G(0, 0) = F_0$, and that the second addend makes a trivial contribution, we obtain

$$F_1 = \tilde{b}(0)\, F_0 * F_0, \tag{30}$$

from which two implications follow: 1) F_1 is determined uniquely by the prescription of F_0, to within a constant factor; 2) Kreĭn's condition (K), formulated in [5], implies that F_0 and consequently also $F_0 * F_0$ are not identically zero, and hence (30) permits finding $\tilde{b}(0)$.

Let us consider $V_2(z)$. By virtue of (28)

$$V_2(z) = \tilde{b}(0)\, G(z, 0) * F_1 - \int [G(z, z_1) * V_1(z_1)]\, \tilde{b}(z_1)\, dz_1 + \int G(z, z_1) * V_0(z_1)\, a(z_1)\, dz_1.$$

Substituting expressions (24) and (29) for $V_0(z_1)$ and $V_1(z_1)$ and using formulas (17) and (20), we obtain

$$V_2(z)=\tilde{b}(0)\,G(z,\ 0)*F_1-\tilde{b}(0)\int dz_1\,[G(z,\ z_1)*G(z_1,\ 0)]*F_0\tilde{b}(z_1)$$

$$-\iint dz_1dz_2\,[G(z,\ z_1)*\varepsilon(z_1-z_2)\,G(z_1,\ 0)]\,\tilde{b}(z_1)\,\beta(z_2)$$

$$+\int dz_1G(z,\ z_1)*G(z_1,\ 0)\,a(z_1)=\tilde{b}(0)\,G(z,\ 0)*F_1$$

$$+\tilde{b}(0)\int dz_1G(z,\ 0)*F_0\varepsilon(z-z_1)\,\beta(z_1)-\iint dz_1dz_2\delta(z_1-z_2)$$

$$\times G(z,\ z_1)*G(z_1,\ 0)\,\tilde{b}(z_1)\,\beta(z_2)+\iint dz_1dz_2G(z,\ 0)\,\varepsilon(z-z_1)\,\varepsilon(z_1-z_2)$$

$$\times\beta(z_1)\,\beta(z_2)+\int dz_1G(z,\ z_1)*G(z_1,\ 0)\,a(z_1)=\tilde{b}(0)\,G(z,\ 0)*F_1$$

$$+\tilde{b}(0)\int dz_1G(z,\ 0)*F_0\varepsilon(z-z_1)\,\beta(z_1)+\iint dz_1dz_2G(z,\ 0)\,\varepsilon(z-z_1)$$

$$\times\beta(z_1)\,\beta(z_2)+\int dz_1G(z,\ z_1)*G(z_1,\ 0)\left(a(z_1)-\frac{\tilde{b}^2(z_1)}{c^2(z_1)}\right). \tag{31}$$

Setting $z=0$ in (31) and introducing the notation

$$a(z)-\frac{\tilde{b}^2(z)}{c^2(z)}\equiv p(z),$$

we obtain

$$F_2=\tilde{b}(0)\,F_0*F_1+\int dz_1G(0,\ z_1)*G(z_1,\ 0)\,p(z_1). \tag{32}$$

Let us show that (32) is a Volterra integral equation of the first kind. It will be convenient for us to utilize the variable $y=\int_0^{z_1}(1/c(\zeta))d\zeta$ instead of z_1 for this purpose. Then equation (32) takes the form

$$F_2=\tilde{b}(0)\,F_0*F_1+\int dyc(y)\,G(0,\ y)*G(y,\ 0)\,p(y).$$

Let us note these properties of the function $G(0,y)=G(y,0)$: 1) $G(0,y)$ is a classical smooth function of the arguments t and y for $t>y$; 2) $G(0,y)=0$ for $t<y$; 3) in a neighborhood of the characteristic $t=y$, $G(0,y)$ has the representation

$$G(0,\ y)=\frac{1}{\sqrt{c(y)\,c(0)}}\,\varepsilon(t-y)+G_1,$$

where G_1 is a continuous function, and therefore

$$G(0,\ y)*G(y,\ 0)=\frac{1}{c(y)\,c(0)}\int\varepsilon(\tau-y)\cdot\varepsilon(t-\tau-y)\,d\tau+G_2$$

$$=\frac{1}{c(y)\,c(0)}\,(t-2y)_++G_2,$$

where G_2 is, by comparison with the terms written out, a smoother function, equal to zero for $t < 2y$.

Now equation (32) reduces to the form

$$F_2(t) - \tilde{b}(0) F_1 * F_0(t) = \frac{2}{c(0)} \int_0^{t/2} \left(\frac{t}{2} - y\right) p(y)\, dy + \int_0^{t/2} G_2(t, y) c(y) p(y)\, dy. \tag{33}$$

It is obvious that after double differentiation with respect to t, equation (33) turns into a linear Volterra integral equation of the second kind.

Thus consideration of the first three moments makes it possible to determine uniquely $c^2(z)$, $\tilde{b}(0)$ and the determinant $D(z)$ of the matrix

$$\begin{bmatrix} a & \tilde{b} \\ \tilde{b} & c^2 \end{bmatrix}.$$

where $D(z) = p(z)c^2(z)$.

§3. Selfadjoint inverse problem. II

The basic result of this section is negative. It consists of proving that even prescription of the whole of the function $f(x,t)$ in (21) does not permit obtaining any other information about the coefficients a, $\tilde{b}$ and c^2 besides that obtained in §2 from the first three moments. It will be proved that each moment $F_k(t)$ of the function $f(x,t)$ for $k = 3, 4, \ldots$ is found explicitly by means of recursion formulas if all the preceding moments $F_\ell(t)$, $\ell < k$, and the functions $p(z)$, $G(z, z_1)$ and $\tilde{b}(0)$ are known. Since the function $f(x,t)$ is a finite function of x for each t, it is found uniquely if all its moments are known. Hence it follows that the function $f(x,t)$ can be reconstructed uniquely if only $c(z)$, $p(z)$ and $\tilde{b}(0)$ are prescribed, independently of the magnitudes $a(z)$ and $\tilde{b}(z)$, from which the converse follows immediately: the functions $a(z)$ and $\tilde{b}(z)$ cannot be found uniquely through $f(x,t)$.

Thus the whole proof reduces to constructing the above-mentioned recursion formulas for the moments $F_k(t)$, to the derivation of which we now proceed.

Let us introduce some new notation. Let

1) $$\Omega(z, z_1, \ldots, z_n) \equiv G(z, z_1) * G(z_1, z_2) * \ldots * G(z_{n-1}, z_n) * G(z_n, 0). \tag{34}$$

2) Let $\chi(z, z_1, \ldots, z_n)$ be a function equal to unity if its arguments are arranged in monotone decreasing order, i.e. $z > z_1 > z_2 > \ldots > z_n$, and equal to zero if even one of these inequalities is violated. Obviously

$$\chi(z, z_1 \ldots z_n) = \varepsilon(z - z_1)\, \varepsilon(z_1 - z_2) \ldots \varepsilon(z_{n-1} - z_n). \tag{35}$$

Let us assume in accordance with the definition that

$$\chi(z) \equiv 1 \text{ and } \Omega(z) \equiv G(z, 0). \tag{36}$$

The equalities

$$G(z, z_1) * \Omega(z_1, \ldots, z_n) = \Omega(z, z_1, \ldots, z_n) \tag{37}$$

and

$$\varepsilon(z - z_1)\chi(z_1, \ldots, z_n) = \chi(z, z_1 \ldots z_n) \tag{38}$$

are obvious.

In the notation (34)–(36) formulas (24), (29) and (31) can be written in the form[4)]

$$V_0(z) = \Omega(z),$$

$$V_1(z) = \tilde{b}(0)\,\Omega(z) * F_0 + \int dz_1 \Omega(z)\chi(z, z_1)\beta(z_1),$$

$$V_2(z) = \tilde{b}(0)\,\Omega(z) * F_1 + \tilde{b}(0)\int dz_1 \Omega(z)\chi(z, z_1)\beta(z_1) * F_0$$

$$+ \int dz_1 dz_2 \Omega(z)\chi(z, z_1, z_2)\beta(z_1)\beta(z_2) + \int dz_1 \Omega(z, z_1)\chi(z_1)p(z_1). \tag{39}$$

It is evident from (39) that the expressions for $V_n(z)$ have the following structure in every case $n = 0, 1, 2$:

$$V_n(z) = \tilde{b}(0)\sum_{k=0}^{n-1} Y_n^k(z) * F_k + Y_n(z), \tag{40}$$

where the functions Y_n and Y_n^k do not depend explicitly on $\tilde{b}(0)$ and F_k. Let us prove (40) by induction for all $n > 0$. Assuming (40) is valid for all V_m, $m < n$, and expressing V_n through V_{n-1} and V_{n-2} in accordance with (28), we find

$$V_n(z) = \tilde{b}(0)\,\Omega(z) * F_{n-1} - \int \left[G(z, z_1) * \left(\tilde{b}(0)\sum_{k=0}^{n-2} Y_{n-1}^k(z_1) * F_k + Y_{n-1}(z_1)\right)\right]$$

$$\times \tilde{b}(z_1)\,dz_1 + \int dz_1 G(z, z_1) * \left(\tilde{b}(0)\sum_{k=0}^{n-3} Y_{n-2}^k(z_1) * F_k + Y_{n-2}(z_1)\right) a(z_1)$$

$$= \tilde{b}(0)\,\Omega(z) * F_{n-1} - \tilde{b}(0)\int [G(z, z_1) * Y_{n-1}^{n-2}(z_1)]\,\tilde{b}(z_1)\,dz_1 * F_{n-2}$$

$$+ \sum_{k=0}^{n-3}\tilde{b}(0)\left\{-\int [G(z, z_1) * Y_{n-1}^k(z_1)]\,\tilde{b}(z_1)\,dz_1 + \int dz_1 G(z, z_1) * Y_{n-2}^k(z_1)\,a(z_1)\right\}$$

$$* F_k - \int [G(z, z_1) * Y_{n-1}(z_1)]\,\tilde{b}(z_1)\,dz_1 + \int dz_1 G(z, z_1) * Y_{n-2}(z_1)\,d(z_1). \tag{41}$$

---

4) In the last of formulas (39) we denoted the double integral with respect to z_1 and z_2 by a single integral sign. In what follows we shall write multiple integrals with a single integral sign, while the number of differentials will indicate the integral's multiplicity.

Formula (40) has been proved, the following recursion formulas being obtained from (41):

$$Y_n^-(z) = -\int dz_1 [G(z, z_1) * Y_{n-1}(z_1)]\tilde{b}(z_1) + \int dz_1 G(z, z_1) * Y_{n-2}(z_1)\, a(z_1), \qquad (42)$$

$$Y_n^k(z) = -\int dz_1 [G(z, z_1) * Y_{n-1}^k(z_1)]\tilde{b}(z_1) + \int dz_1 G(z, z_1) * Y_{n-2}^k(z_1)\, a(z_1). \qquad (43)$$

Formulas (42) and (43) are valid respectively for $n > 1$ and $n > k + 2$; they prove to be valid for $n = 1$ and $n = k + 2$ if we correspondingly set $Y_{-1} = Y_k^k \equiv 0$. Let us observe that the first formula of (39) and (41) imply that

$$Y_0 = Y_{k+1}^k = \Omega(z). \qquad (44)$$

But since recursion formulas (42) and (43) coincide, we obtain $Y_m = Y_{k+m+1}^k$, formula (40) acquires the form

$$V_n(z) = \tilde{b}(0) \sum_{k=0}^{n-1} Y_{n-k-1}(z) * F_k + Y_n(z), \qquad (45)$$

and consequently we need evaluate only $Y_n(z)$.

Let us show that

$$Y_n(z) = \sum_{l=0}^{[n/2]} X_{n,l}(z), \qquad (46)$$

where $[n/2]$ is the integral part of $n/2$ and

$$X_{nl}(z) = \int dz_1 \ldots dz_{n-l}\, \Omega(z, z_1 \ldots z_l) \times \chi(z, z_{l+1}, \ldots, z_{n-l})\, p(z_1) \ldots p(z_l)\, \beta(z_{l+1}) \ldots \beta(z_{n-l}) \qquad (47)$$

for $1 \leqslant [n/2]$, $n \geqslant 1$.

Before proving (46), let us prove that it and (45) imply that F_n is determined uniquely if all F_k for $k < n$ are known, as well as $\tilde{b}(0)$, $p(z)$ and $G(z, z_1)$, independently of the values $\beta(z)$ for $z > 0$ [of course this result is true only for the moments F_n of the boundary value of the function $u(x, z, t)$ and not for the moments $V_n(z)$ of the function $u(x, z, t)$, itself]. Actually (47) and the properties of the function χ imply immediately that

$$X_{n,l}(0) = 0$$

for $l < n - l$, i.e. for $l < n/2$. Therefore

$$Y_n(0) = \begin{cases} 0 & \text{for odd } n \\ X_{n,\frac{n}{2}}(0) & \text{for even } n \end{cases} \tag{48}$$

which implies that F_n, which by virtue of (45) equals

$$F_n = \tilde{b}(0) \sum_{k=0}^{n-1} Y_{n-k-1}(0) * F_k + Y_n(0), \tag{49}$$

is expressed only through $\tilde{b}(0)$, F_k and $X_{2l,l}(0)$, where $k < n$ and $l \leqslant n$. But, according to (47),

$$X_{2l,l}(0) = \int dz_1 \ldots dz_l \Omega(0, z_1, \ldots, z_l) p(z_1) \ldots p(z_l)$$

$$= \int dz_1 \ldots dz_l G(0, z_1) * G(z_1, z_2) * \ldots * G(z_l; 0) p(z_1) \ldots p(z_l),$$

which proves the assertion made above.

Let us pass to the proof of (46). It is clear from (39) that (46) is valid for $n = 1$ and 2. In order to carry out induction, let us first make an evaluation of integrals of the form

$$-\int dz_1 [G(z, z_1) * X_{n-1,l}(z_1)] \tilde{b}(z_1)$$

$$= -\int dz_1 \ldots dz_{n-l} [G(z, z_1) * \varepsilon(z_1 - z_{l+2}) G(z_1, z_2)]$$

$$* \Omega(z_2, \ldots, z_{l+1}) \chi(z_{l+2}, z_{l+3}, \ldots, z_{n-l}) \tilde{b}(z_1) p(z_2) \ldots p(z_{l+1})$$

$$\times \beta(z_{l+2}) \ldots \beta(z_{n-l}) \quad \text{for} \quad 0 < l < \frac{n-1}{2}.$$

Substituting expression (20) for the convolution Wronskian, we obtain

$$-\int dz_1 [G(z, z_1) * X_{n-1,l}(z_1)] \tilde{b}(z_1)$$

$$= \int dz_1 \ldots dz_{n-l} G(z, z_2) * \Omega(z_2, \ldots, z_{l+1}) \varepsilon(z_1 - z_{l+2})$$

$$\times [\varepsilon(z_1 - z_2) - \varepsilon(z_1 - z)] \chi(z_{l+2}, z_{l+3}, \ldots, z_{n-l}) \beta(z_1)$$

$$\times p(z_2) \ldots p(z_{l+1}) \beta(z_{l+2}) \beta(z_{l+3}) \ldots \beta(z_{n-l})$$

$$-\int dz_1 \ldots dz_{n-l} \delta(z_1 - z_{l+2}) G(z, z_1) * G(z_1, z_2) * \Omega(z_2, \ldots, z_{l+1})$$

$$\times \chi(z_{l+2}, z_{l+3}, \ldots, z_{n-l}) \tilde{b}(z_1) p(z_2) \ldots p(z_{l+1}) \beta(z_{l+2}) \ldots \beta(z_{n-l}).$$

Reducing the multiplicity of the integral in the second addend at the expense of evaluating the integral from the δ-function, effecting transformations of the variables of integration, utilizing the formula $\tilde{b}(z_1)\beta(z_1) = \tilde{b}^2(z_1)/c^2(z_1)$ and applying (37) and (38), we find

$$
\begin{aligned}
&-\int dz_1\,[G(z,\ z_1)*X_{n-1,l}(z_1)]\,\tilde{b}(z_1)\\
&\quad=\int dz_1\ldots dz_{n-l}\,\Omega(z,\ z_1,\ldots,z_l)\,[\varepsilon(z_{l+1}-z_1)-\varepsilon(z_{l+1}-z)]\\
&\quad\times\chi(z_{l+1},\ z_{l+2},\ldots,z_{n-l})\,p(z_1)\ldots p(z_l)\,\beta(z_{l+1})\ldots\beta(z_{n-l})\\
&\quad-\int dz_1\ldots dz_{n-l-1}\,\Omega(z,\ z_1,\ldots,z_{l+1})\,\chi(z_1,\ z_{l+2},\ldots,z_{n-l-1})\\
&\qquad\times\frac{\tilde{b}^2(z_1)}{c^2(z_1)}\,p(z_2)\ldots p(z_{l+1})\,\beta(z_{l+2})\ldots\beta(z_{n-l-1}).
\end{aligned}
\tag{50}
$$

As we mentioned above, (50) is valid for $0 < l < (n-1)/2$. The formula

$$
\begin{aligned}
&-\int dz_1\,[G(z,\ z_1)*X_{n-1,0}(z_1)]\,\tilde{b}(z_1)\\
&\quad=-\int dz_1\ldots dz_n\,[G(z,\ z_1)*\varepsilon(z_1-z_2)\,G(z_1,\ 0)]\\
&\qquad\times\chi(z_2,\ldots,z_n)\,\tilde{b}(z_1)\,\beta(z_2)\ldots\beta(z_n)\\
&\quad=\int dz_1\ldots dz_n\,\Omega(z)\,\chi(z,\ z_1,\ldots,z_n)\,\beta(z_1)\ldots\beta(z_n)\\
&\qquad-\int dz_1\ldots dz_{n-1}\,\Omega(z,\ z_1)\,\chi(z_1,\ldots,z_{n-1})\,\frac{\tilde{b}^2(z_1)}{c^2(z_1)}\,\beta(z_2)\ldots\beta(z_{n-1})\\
&\quad=X_{n0}(z)-\int dz_1\ldots dz_{n-1}\,\Omega(z,\ z_1)\,\chi(z_1,\ldots,z_{n-1})\,\frac{\tilde{b}^2(z_1)}{c^2(z_1)}\,\beta(z_2)\ldots\beta(z_{n-1}).
\end{aligned}
\tag{51}
$$

is valid for $l = 0$. For odd n and $l = (n-1)/2$

$$
\begin{aligned}
&-\int dz_1\left[G(z,\ z_1)*X_{n-1,\frac{n-1}{2}}(z_1)\right]\tilde{b}(z_1)\\
&\quad=-\int dz_1\ldots dz_{\frac{n+1}{2}}\,[G(z,\ z_1)*G(z_1,\ z_2)]*\Omega\left(z_2,\ldots,z_{\frac{n+1}{2}}\right)\\
&\times\chi(z_1)\,p(z_2)\ldots p\left(z_{\frac{n+1}{2}}\right)\tilde{b}(z_1)=\int dz_1\ldots dz_{\frac{n+1}{2}}\,\Omega\left(z,\ z_1,\ldots,z_{\frac{n-1}{2}}\right)\\
&\qquad\times\left[\varepsilon\left(z_{\frac{n+1}{2}}-z_1\right)-\varepsilon\left(z_{\frac{n+1}{2}}-z\right)\right]p(z_1)\ldots p\left(z_{\frac{n-1}{2}}\right)\beta\left(z_{\frac{n+1}{2}}\right).
\end{aligned}
\tag{52}
$$

We used the same steps for the derivation of (51) and (52) as for the derivation of (50). Now let (46) be established for all $Y_m(z)$, where $m < n$. We show that this hypothesis implies the validity of (46) for $Y_n(z)$, also. Let us examine the two cases separately.

1. Let n be even. Then utilizing recursion formula (42) and substituting (46), which is valid by the induction hypothesis, for $Y_{n-1}(z_1)$ and $Y_{n-2}(z_1)$ in (42), we obtain

$$Y_n(z) = -\int dz_1 [G(z, z_1) * Y_{n-1}(z_1)] \tilde{b}(z_1)$$

$$+ \int dz_1 G(z, z_1) * Y_{n-2}(z_1) a(z_1) = \sum_{l=0}^{n/2-1} - \int dz_1 [G(z, z_1) * X_{n-1,l}(z_1)] \tilde{b}(z_1)$$

$$+ \sum_{l=0}^{n/2-1} \int dz_1 G(z, z_1) * X_{n-2,l}(z_1) a(z_1).$$

Applying (50), (51) and (47), we obtain

$$Y_n(z) = X_{n,0}(z) - \int dz_1 \dots dz_{n-1} \Omega(z, z_1) \chi(z_1, \dots, z_{n-1})$$

$$\times \frac{\tilde{b}^2(z_1)}{c^2(z_1)} \beta(z_2) \dots \beta(z_{n-1}) + \sum_{l=1}^{n/2-1} \int dz_1 \dots dz_{n-l} \Omega(z, z_1, \dots, z_l)$$

$$\times [\varepsilon(z_{l+1} - z_1) - \varepsilon(z_{l+1} - z)] \chi(z_{l+1}, z_{l+2}, \dots, z_{n-l}) p(z_1) \dots p(z_l) \beta(z_{l+1})$$

$$\dots \beta(z_{n-l}) - \sum_{l=1}^{n/2-1} \int dz_1 \dots dz_{n-l-1} \Omega(z, z_1 \dots z_{l+1}) \chi(z_1, z_{l+2}, \dots, z_{n-l-1})$$

$$\times \frac{\tilde{b}^2(z_1)}{c^2(z_1)} p(z_2) \dots p(z_{l+1}) \beta(z_{l+2}) \dots \beta(z_{n-l-1}) + \sum_{l=0}^{n/2} \int dz_1 \dots dz_{n-l-1}$$

$$\times \Omega(z, z_1, \dots, z_{l+1}) \chi(z_1, z_{l+2}, \dots, z_{n-l-1}) a(z_1) p(z_2) \dots$$

$$p(z_{l+1}) \beta(z_{l+2}) \dots \beta(z_{n-l-1}).$$

Recalling that $a(z_1) - \tilde{b}^2(z_1)/c^2(z_1) = p(z_1)$ and making insignificant regroupings of the terms, we find

$$Y_n(z) = X_{n0}(z) + \sum_{l=0}^{n/2-2} \int dz_1 dz_2 \dots dz_{n-l-1} \Omega(z, z_1, \dots, z_{l+1})$$

$$\times \chi(z_1, z_{l+2}, \dots, z_{n-l-1}) p(z_1) \dots p(z_{l+1}) \beta(z_{l+2}) \dots \beta(z_{n-l-1})$$

$$+ \sum_{l=1}^{n/2-1} \int dz_1 \dots dz_{n-l} \Omega(z, z_1, \dots, z_l) [\varepsilon(z_{l+1} - z_1) - \varepsilon(z_{l+1} - z)]$$

$$\times \chi(z_{l+1}, z_{l+2}, \dots, z_{n-l}) p(z_1) \dots p(z_l) \beta(z_{l+1}) \dots \beta(z_{n-l})$$

$$+ \int dz_1 \dots dz_{n/2} \Omega(z, z_1, \dots, z_{n/2}) \chi(z_1) p(z_1) \dots p(z_{n/2}).$$

Changing the index of summation l to $l - 1$ in the first sum, consolidating it with the second sum, and observing that the latter term is $X_{n,n/2}(z)$, we have

$$Y_n(z) = X_{n0}(z) + \sum_{l=1}^{n/2-1} \int dz_1 \dots dz_{n-l} \Omega(z, z_1 \dots z_l)$$
$$\times [\varepsilon(z_{l+1} - z_1) - \varepsilon(z_{l+1} - z) + \varepsilon(z_1 - z_{l+1})] \chi(z_{l+1}, \dots, z_{n-l})$$
$$\times p(z_1) \dots p(z_l) \beta(z_{l+1}) \dots \beta(z_{n-l}) + X_{n, n/2}(z) = X_{n,0}(z)$$
$$+ \sum_{l=1}^{n/2-1} \int dz_1 \dots dz_{n-l} \Omega(z, z_1, \dots, z_l) \chi(z, z_{l+1}, \dots, z_{n-l})$$
$$\times p(z_1) \dots p(z_l) \beta(z_{l+1}) \dots \beta(z_{n-l}) + X_{n, n/2}(z) = \sum_{l=0}^{n/2} X_{n, l}(z).$$

Formula (46) is proved.

2. Let n be odd. Since the transformations carried out here are very similar to those realized in the case of even n, we shall perform them here without additional explanations:

$$Y_n(z) = \sum_{l=0}^{(n-1)/2} - \int dz_1 [G(z, z_1) * X_{n-1, l}(z_1)] \tilde{b}(z_1)$$
$$+ \sum_{l=0}^{(n-3/2)} \int dz_1 G(z, z_1) * X_{n-2, l}(z_1) a(z_1) = X_{n0}(z)$$
$$+ \sum_{l=1}^{(n-1)/2} \int dz_1 \dots dz_{n-l} \Omega(z, z_1, \dots, z_l) [\varepsilon(z_{l+1} - z_1) - \varepsilon(z_{l+1} - z)]$$
$$\times \chi(z_{l+1}, z_{l+2}, \dots, z_{n-l}) p(z_1) \dots p(z_l) \beta(z_{l+1}) \dots \beta(z_{n-l}) -$$
$$- \sum_{l=0}^{(n-3)/2} \int dz_1 \dots dz_{n-l-1} \Omega(z, z_1, \dots, z_{l+1}) \chi(z_1, z_{l+2}, \dots, z_{n-l-1})$$
$$\times \frac{\tilde{b}^2(z_1)}{c^2(z_1)} p(z_2) \dots p(z_{l+1}) \beta(z_{l+2}) \dots \beta(z_{n-l-1}) +$$
$$+ \sum_{l=0}^{(n-1),2} \int dz_1 \dots dz_{n-l-1} \Omega(z, z_1, \dots, z_{l+1}) \chi(z_1, z_{l+2}, \dots, z_{n-l-1})$$
$$\times a(z_1) p(z_2) \dots p(z_{l+1}) \beta(z_{l+2}) \dots \beta(z_{n-l-1}) = X_{n,0}(z) +$$
$$+ \sum_{l=1}^{(n-1)/2} \int dz_1 \dots dz_{n-l} \Omega(z, z_1, \dots, z_l) [\varepsilon(z_{l+1} - z_1) - \varepsilon(z_{l+1} - z)]$$
$$\times \chi(z_{l+1}, \dots, z_{n-l}) p(z_1) \dots p(z_l) \beta(z_{l+1}) \dots \beta(z_{n-l}) +$$
$$+ \sum_{l=1}^{(n-1)/2} \int dz_1 \dots dz_{n-l} \Omega(z, z_1, \dots, z_l) \varepsilon(z_1 - z_{l+1}) \chi(z_{l+1}, \dots, z_{n-l})$$
$$\times p(z_1) \dots p(z_l) \beta(z_{l+1}) \dots \beta(z_{n-l}) = X_{n0}(z) + \sum_{l=1}^{(n-1)/2} \int dz_1 \dots dz_{n-l}$$
$$\times \Omega(z, z_1, \dots, z_l) \chi(z, z_{l+1}, \dots, z_{n-l}) p(z_1) \dots p(z_l) \beta(z_{l+1}) \dots \beta(z_{n-l})$$
$$= \sum_{l=0}^{(n-1)/2} X_{nl}(z) = \sum_{l=0}^{[n/2]} X_{nl}(z).$$

Formula (46) is completely proved.

§4. Nonselfadjoint problem

In this section we consider the problem of determining the coefficients $a(z)$, $b(z)$ and $c(z)$ in the hyperbolic equation

$$u_{tt}=c^2u_{zz}+2bcu_{zx}+au_{xx} \tag{53}$$

under the conditions

$$u|_{t<0}=0, \quad u_z|_{z=0}=\delta(x, t), \tag{54}$$

$$\frac{1}{k!}\int u(x, 0, t)\,x^k dx=F_k(t) \qquad (k=0, 1, 2), \tag{55}$$

where the $F_k(t)$ are sufficiently smooth prescribed functions. By comparison with formula (5) we have here changed the notation somewhat by introducing, instead of $\tilde{b}(z)$, the unknown function $b(z) : \tilde{b}(z) - b(z)c(z)$.

Let us define the moments of the function $u(x, z, t)$ by

$$V_k(z, t)=\frac{1}{k!}\int u(x, z, t)\,x^k dx \tag{56}$$

From (53)–(55) we conclude that $V_0(z, t)$ is a solution of the following problem:

$$\left.\begin{aligned} &V_{0tt}=c^2V_{0zz}, \\ &V_{0z}|_{z=0}=\delta(t), \quad V_0|_{z=0}=F_0(t) \end{aligned}\right\} \tag{57}$$

(the condition $V_k|_{t<0}=0$ is implicit everywhere below).

Problem (57) has been analyzed, in [5] for example. Here we assume that it has already been solved and the functions $c(z)$ and $V_0(z, t)$ found.

Let us write out a problem for $V_1(z, t)$:

$$\left.\begin{aligned} &V_{1tt}=c^2V_{1zz}-2bcV_{0z}, \\ &V_1|_{z=0}=F_1(t), \quad V_{1z}|_{z=0}=0. \end{aligned}\right\} \tag{58}$$

Let us make the standard change of variables $y = \int_0^z (1/c(\zeta))d\zeta$. Then we arrive at the problem

$$\left.\begin{aligned} &V_{1tt}=V_{1yy}-\frac{c'}{c}V_{1y}-2bV_{0y}, \\ &V_1|_{y=0}=F_1(t), \quad V_{1y}|_{y=0}=0. \end{aligned}\right\} \tag{59}$$

Let $H(y, \eta, t)$ and $K(y, t)$ be solutions of the problems

$$\left.\begin{aligned} H_{yy}-\frac{c'}{c}H_y-H_{tt}&=\delta(t)\,\delta(y-\eta),\\ H\,|_{y<\eta}&\equiv 0;\end{aligned}\right\} \tag{60}$$

$$\left.\begin{aligned} K_{yy}-\frac{c'}{c}K_y-K_{tt}&=0,\\ K\,|_{y=0}=\delta(t),\quad K_y\,|_{y=0}&=0.\end{aligned}\right\} \tag{61}$$

Then it is easy to obtain from (59) that

$$V_1(y,t)=2\iint H(y,\eta,t-\tau)\,b(\eta)\,V_{0\eta}(\eta,\tau)\,d\eta d\tau+\int K(y,t-\tau)\,F_1(\tau)\,d\tau. \tag{62}$$

We find from the theory of characteristics that in a neighborhood of the straight line $t=y$

$$V_0=-\sqrt{c(y)\,c(0)}\,\varepsilon(t-y)+\dots, \tag{63}$$

$$V_1=P_1(y)\,\varepsilon(t-y)+\dots \tag{64}$$

The addends which are less singular on the characteristic are denoted by dots. The function $P_1(y)$ satisfies the differential equation

$$-2P_1'+\frac{c'}{c}P_1=2\sqrt{c(y)\,c(0)}\,b(y)$$

or

$$b(y)=-\frac{1}{\sqrt{c(0)}}\left(\frac{P_1(y)}{\sqrt{c(y)}}\right)_y'. \tag{65}$$

Using formula (62) and the fact that $P_1(y)=V_1(y,y)$, we obtain from (65)

$$b(y)=-\frac{1}{\sqrt{c(0)}}\frac{d}{dy}\left[\frac{2}{\sqrt{c(y)}}\iint H(y,\eta,y-\tau)\,b(\eta)\,V_{0\eta}(\eta,\tau)\,d\eta d\tau\right.$$
$$\left.+\frac{1}{\sqrt{c(y)}}\int K(y,y-\tau)\,F_1(\tau)\,d\tau\right]. \tag{66}$$

Equation (66) is a linear Volterra integral equation of the second kind relative to the function $b(y)$, with kernel

$$T_1(y,\eta)=\frac{\partial}{\partial y}2\int\frac{H(y,\eta,y-\tau)}{\sqrt{c(y)}}V_{0\eta}(\eta,\tau)\,d\tau.$$

It is possible to find $b(y)$ from this equation. Let us assume that this has been done and $b(y)$ as well as $V_1(t,y)$ have been found.

Let us consider a problem for V_2. From (53)–(55) it follows that

$$\left.\begin{aligned} V_{2tt} &= V_{2yy} - \frac{c'}{c} V_{2y} - 2bV_{1y} + aV_0, \\ V_2\big|_{y=0} &= F_2(t), \quad V_{2y}\big|_{y=0} = 0. \end{aligned}\right\} \tag{67}$$

Quite analogously to the foregoing,

$$V_2(y, t) = -\iint H(y, \eta, t-\tau) V_0(\eta, \tau) a(\eta) \, d\eta d\tau + \Phi(y, t), \tag{68}$$

where $\Phi(y, t)$ is a known function equal to

$$\Phi(y, t) = 2\iint H(y, \eta, t-\tau) V_{1\eta}(\eta, \tau) b(\eta) \, d\eta d\tau + \int K(y, t-\tau) F_2(\tau) \, d\tau.$$

In a neighborhood of the characteristics $t = y$ the function $V_2(y, t)$ has the form

$$V_2(y, t) = P_2(y) \varepsilon(t-y) + \ldots, \tag{69}$$

where $P_2(y)$ satisfies the differential equation

$$2P_2' - \frac{c'}{c} P_2 = 2\left(\frac{P_2}{\sqrt{c}}\right)' \sqrt{c} = 2bP_1; \tag{70}$$

the function $b(y)P_1(y) = b(y)V_1(y, y)$ is known.

Substituting $V_2(y, y)$ for $P_2(y)$ in (70) and utilizing (68), we obtain

$$\left(\frac{P_2}{\sqrt{c}}\right)' = -\frac{d}{dy}\iint \frac{1}{\sqrt{c(y)}} H(y, \eta, y-\tau) V_0(\eta, \tau) a(\eta) \, d\eta d\tau + \frac{d}{dy}\frac{\Phi(y, y)}{\sqrt{c(y)}} = b(y) P_1(y) \frac{1}{\sqrt{c(y)}}. \tag{71}$$

(71) is a Volterra integral equation of the first kind relatively to $a(y)$, with kernel

$$T_2(y, \eta) = -\frac{d}{dy}\frac{1}{\sqrt{c(y)}}\int H(y, \eta, y-\tau) V_0(\eta, \tau) \, d\tau. \tag{72}$$

Obviously the kernel $T_2(y, \eta) \equiv 0$ for $\eta > y$, since $H(y, \eta, t) \equiv 0$ for $\eta > y$. Let us investigate the behavior of the kernel $T_2(y, \eta)$ in a neighborhood of a point of $y = \eta$ for $y > \eta$.

Let us consider the function

$$\frac{1}{\sqrt{c(y)}}\int H(y, \eta, y-\tau) V_0(\eta, \tau) \, d\tau.$$

It is easy to show that $H(y, \eta, t)$ has the representation

$$H(y, \eta, t) = \frac{1}{2}\varepsilon(y-\eta-t)\,\varepsilon(t+y-\eta)\,(1+O(y-\eta)).$$

in a neighborhood of a point of $y - \eta = t = 0$. $V_0(\eta, \tau)$ has the form

$$V_0 = -\sqrt{c(y)\,c(0)} + O(y-\eta+|\tau-y|).$$

in a neighborhood of a point of $\eta = y$, $\tau = y$ for $\tau > \eta$, $y > \eta$. Therefore

$$\frac{1}{\sqrt{c(y)}}\int H(y, \eta, y-\tau)\,V_0(\eta, \tau)\,d\tau = \frac{1}{\sqrt{c(y)}}\int_{\eta}^{2y-\eta} H(y, \eta, y-\tau)\,V_0(\eta, \tau)\,d\tau$$

$$= -\frac{\sqrt{c(0)}}{2}\int_{\eta}^{2y-\eta} d\tau + O((y-\eta)^2) = -\sqrt{c(0)}\,(y-\eta) + O((y-\eta)^2),$$

whence

$$T_2(y, \eta) = -\sqrt{c(0)}\,\varepsilon(y-\eta) + O(y-\eta). \tag{73}$$

Consequently, after differentiation with respect to y, equation (71) turns into a Volterra integral equation of the second kind.

Thus, finally, if $F_0(t)$ satisfies the condition (K) of [5], the coefficients a, b and c in the nonselfadjoint problem (53)–(55) can be determined uniquely for all $z > 0$, in contrast with the selfadjoint case.

BIBLIOGRAPHY

1. A.S. Blagoveščenskiĭ, *On an inverse problem in the theory of seismic wave distribution*, Problemy Mat. Fiz., no. 1, Izdat. Leningrad. Univ., Leningrad, 1966, pp. 68–81; English transl., Topics in Math. Phys., no. 1, Consultants Bureau, New York, 1967, pp. 55–67.
2. A.A. Buzdin, *On an inverse problem in a medium with absorption*, Trudy Mat. Inst. Steklov. **115** (1971), 70–77 = Proc. Steklov Inst. Math. **115** (1971), 77–85.
3. A.S. Blagoveščenskiĭ, *On a quasi-two-dimensional inverse problem for a wave equation*, Trudy Mat. Inst. Steklov. **115** (1971), 57–69 = Proc. Steklov Inst. Math. **115** (1971), 63–76.
4. I.M. Gel'fand and G.E. Šilov, *Generalized functions*. Vol. 1: *Operations on them*, Fizmatgiz, Moscow, 1958; English transl., Academic Press, New York, 1964. MR **20** #4182; **29** #3869.
5. A.S. Blagoveščenskiĭ, *On a local method of solving a nonstationary inverse problem for an inhomogeneous string*, Trudy Mat. Inst. Steklov. **115** (1971), 28–38 = Proc. Steklov Inst. Math. **115** (1971), 29–40.
6. M.G. Kreĭn, *On a method of effective solution of an inverse problem*, Dokl. Akad. Nauk SSSR **94** (1954), 987–990. (Russian) MR **16**, 38.

Translated by:
F.M. Goldware

A QUASI-TWO-DIMENSIONAL PROBLEM FOR THE WAVE EQUATION

A. S. BLAGOVEŠČENSKIĬ

UDC 517 944/947

Abstract. The article considers the problem of finding the coefficient q in the equation

$$u_{tt} = u_{xx} + u_{zz} + q(x, z)\,u \quad (z>0)$$

from certain rather cumbersomely formulated characteristics of the function $f(x,t) \equiv u(x,z,t)|_{z=0}$. It is assumed that q has the form

$$q(\alpha x, z) = q_0(z) + \alpha x q_1(z) + \dots + \alpha^n x^n q_n(z) + 0(\alpha^n),$$

where α is some "small" parameter. The problem reduces to finding the coefficients of the expansion; this in turn reduces to the solution of a number of one-dimensional problems.

Bibliography: 7 items.

The problem dealt with in this article is that of finding the unknown function $q(x,z)$ which appears in the equation

$$u_{tt} = u_{xx} + u_{zz} + q(x, z)\,u \quad (z > 0), \tag{I}$$

subject to the condition that

$$u|_{t<0} \equiv 0, \quad u_z|_{z=0} = \delta(x, t), \tag{II}$$

where $\delta(x,t)$ is Dirac's delta function and we are given certain information about the function $f(x,t)$, where $f(x,t)$ is the value of $u(x,z,t)|_{z=0}$. It is assumed that the dependence of $q(x,z)$ on x is weak and highly specific. More precisely, we will assume that $q(x,z) = q(\alpha x, z)$, where α is a small parameter. The function $q(\alpha x, z)$ will be assumed to be smooth, so that

$$q(\alpha x, z) = q_0(z) + \alpha x q_1(z) + \dots + \alpha^n x^n q_n(z) + o(\alpha^n).$$

The problem is to find the coefficients $q_i(z)$. This problem will be called quasi-two-dimensional: the unknown function q is a function of two arguments, but because this dependence is specific and weak the problem reduces (as we will see below) to the solution of a sequence of one-dimensional problems.

To indicate precisely what is assumed to be given in our problem, we will perform some preliminary operations. It is easy to show that u and f as functions of α have the structure

$$u(x, z, t) = u_0(x, z, t) + \alpha u_1(x, z, t) + \dots + \alpha^n u_n(x, z, t) + o(\alpha^n),$$

$$f(x, t) = f_0(x, t) + \alpha f_1(x, t) + \dots + \alpha^n f_n(x, t) + o(\alpha^n).$$

AMS 1970 *subject classifications*. Primary 35L05, 35R30; Secondary 44A35, 44A50, 45D05.

By using standard calculations, substituting the expansions of q and u in powers of α into (I) and equating coefficients for equal powers of α, we readily obtain differential equations for the functions $u_k(x,t)$:

$$u_{ktt} = u_{kxx} + u_{kzz} + \sum_{j=0}^{k} q_j(z) x^j u_{k-j}.$$

Multiplying this equation by x^m and integrating with respect to x from minus to plus infinity, we obtain

$$V_{k,mtt} = V_{k,mzz} + m(m-1) V_{k,m-2} + \sum_{j=0}^{k} q_j(z) V_{k-j,m+j}. \tag{1}$$

In (1) we have used $V_{k,m}$ to denote the mth moment of u_k :

$$V_{k,m}(z,t) \equiv \int_{-\infty}^{\infty} u_k(x,z,t) x^m dx.$$

It is obvious that all the $V_{k,m}$ satisfy the conditions

$$V_{km}|_{t<0} \equiv 0, \tag{2}$$

$$V_{kmz}|_{z=0} = \delta_{k0}\delta_{m0}\delta(t), \tag{3}$$

where $\delta_{k0}\,\delta_{m0}$ are Kronecker deltas.

Let

$$V_{k\sigma_k}|_{z=0} = \int_{-\infty}^{\infty} f_k(x,t) x^{\sigma_k} dx \equiv a_k(t), \tag{4}$$

where $\sigma_k = 1$ for odd k and 0 for even k.

We will assume that the functions $a_0(t)$ and $b_k(t)$, $k = 1, \ldots, n$, are given, where

$$b_k(t) \equiv \frac{1}{(k+\sigma_k-1)!!} \frac{\partial^{\frac{k+\sigma_k}{2}+1}}{\partial t^{\frac{k+\sigma_k}{2}+1}} a_k(t) = \frac{1}{(k+\sigma_k-1)!!} \frac{\partial^{\frac{k+\sigma_k}{2}+1}}{\partial t^{\frac{k+\sigma_k}{2}+1}} V_{k\sigma_k}|_{z=0}. \tag{5}$$

The functions $a_0(t)$ and $b_k(t)$ are assumed to be smooth and to satisfy the condition

$$b_k(t) = O\left(t^{\frac{k+\sigma_k}{2}+1}\right) \tag{6}$$

as $t \to 0$.

REMARK. As formulated, problem (1)–(6) is not a physical problem. The functions $a_k(t)$ are coefficients in the expansion of the zero and first moments of $f(x,t)$:

$$\left.\begin{aligned}\int f(x,t)\,dx &= a_0(t)+\alpha^2 a_2(t)+\alpha^4 a_4(t)+\dots,\\ \int f(x,t)\,x\,dx &= \alpha a_1(t)+\alpha^3 a_3(t)+\alpha^5 a_5(t)+\dots\end{aligned}\right\} \qquad (*)$$

(the coefficients for odd powers of α in the first formula in $(*)$ and for the even powers in the second formula are equal to 0, as is shown in Lemma 1 below). To find the $a_k(t)$, therefore, it is necessary to know the function $f(x,t)$ as a function of α, and hence for a real physical problem we would require a series of experiments involving a set of different media corresponding to different values of α; in a real situation we usually deal with the problem of finding the properties of some fixed medium. Nonetheless, formulas $(*)$ imply that the first two moments of f may be used to determine $a_0(t)$ to within a term of order $O(\alpha^2)$ and $\alpha a_1(t)$ to within $O(\alpha^3)$; as is shown below, this makes it possible to calculate $q_0(z)$ and $q_1(z)$ correct to order $O(\alpha^2)$.

We set $k = m = 0$. Then (1)–(6) leads to the problem

$$\left.\begin{aligned}&V_{00tt}=V_{00zz}+q_0(z)V_{00},\\ &V_{00}|_{t<0}\equiv 0,\quad V_{00z}|_{z=0}=\delta(t),\quad V_{00}|_{z=0}=a_0(t).\end{aligned}\right\} \qquad (7)$$

Problem (7) (the function $q_0(z)$ is assumed to be unknown) was studied in detail in [1–4]. It has been shown that problem (7) will be solvable if and only if the quadratic form

$$A(\varphi,\varphi)=-\int_0^\infty\int_0^\infty dt_1dt_2\,[a_0'(|t_1-t_2|)+a_0'(t_1+t_2)]\,\varphi(t_1)\,\varphi(t_2)+2\int_0^\infty|\varphi(t)|^2\,dt$$

is positive definite. We will assume that this condition is satisfied and that problem (7) is solved. Our main interest in this paper will be to determine the subsequent terms $q_i(z)$.

Our basic result may be formulated as follows. Problem (1)–(6), after problem (7) has been solved, reduces to the successive solution of a chain of correct problems; the "links" in this chain are either direct problems (mixed problems for hyperbolic equations) or linear inverse problems whose solution will be reduced to the solution of Volterra integral equations of the second kind.

We will now prove the above results. First we will prove the following lemma.

LEMMA I. *Assume that $q_0,\dots,q_k$ are known, and that the functions V_{js} are known for all $0\leqslant j<k$ and all $0\leqslant s<\infty$, and that for $j=k$ they are known for $s<m$. Then* 1) *a unique function V_{km} exists*; 2) *for $k+m$ odd all the $V_{mk}\equiv 0$.*

For V_{km} we have the direct problem

$$\left.\begin{aligned} V_{kmtt} = V_{kmzz} + q_0(z) V_{km} + m(m-1) V_{km-2} \\ + \sum_{j=1}^{k-1} q_j V_{k-jm+j} + q_k(z) V_{0m+k}, \\ V_{kmz}|_{z=0} = 0 \end{aligned}\right\} \tag{8}$$

(the zero initial condition $V_{km}|_{t<0} \equiv 0$ is always understood in this article, as in [5, 6]). More briefly, problem (8) may be written as

$$\left.\begin{aligned} V_{kmtt} = V_{kmzz} + q_0 V_{km} + \Phi_{km}, \\ V_{kmz}|_{z=0} = 0, \end{aligned}\right\}$$

where Φ_{km} is a function which is known in view of the conditions of the lemma, and hence the first assertion of the lemma is trivial.

In particular, for $k = 0$ and $m = 1$ expression (8) yields

$$\left.\begin{aligned} V_{01tt} = V_{01zz} + q_0(z) V_{01}, \\ V_{01z}|_{z=0} = 0, \end{aligned}\right\}$$

whence it is obvious that $V_{01} \equiv 0$ for all t. Now we note that Φ_{km} depends only on those V_{js} for which the parity of the numbers $j + s$ and $k + m$ is the same; the nature of the dependence is linear. By induction, the last remark and the fact that $V_{01} \equiv 0$ readily yield that all the V_{km} for odd $k + m$ are equal to 0. Lemma I is thus proved.

The inverse problem on finding all the $q_k(z)$ will be solved inductively. We will assume that $q_0(z), \ldots, q_{k-1}(z)$ have been determined and that all the V_{js} for $j < k$ are also known. We introduce the functions

$$W_{km} \equiv \frac{1}{(k+m-1)!!} \frac{\partial^{\frac{k+m}{2}+1}}{\partial t^{\frac{k+m}{2}+1}} V_{km}. \tag{9}$$

Then for $W_{k\sigma_k}$ and q_k we have the problem

$$W_{k\sigma_k tt} = W_{k\sigma_k zz} + q_0(z) W_{k\sigma_k} + q_k(z) W_{0\sigma_k+k} + \Psi_k(z, t), \tag{10}$$

where

$$\Psi_k(z, t) \equiv \sum_{j=1}^{k-1} q_j W_{k-j\sigma_k+j} \tag{11}$$

is a known function in view of the inductive hypothesis;

$$W_{k\sigma_k}|_{z=0} = b_k(t), \tag{12}$$

$$W_{k\sigma_k z}|_{k=0} = 0. \tag{13}$$

It will be important for us to study the singularity of $W_{k\sigma_k}$ on the characteristic $t = z$ of (10) (henceforth this will simply be called the characteristic). For this it is necessary to study the behavior of the W_{js} on the characteristic. We will also need to study the function $W_{0k+\sigma_k}$ in the neighborhood of the coordinate origin. The results of these investigations are incorporated into the following lemmas, which are proved under the assumption that the inverse problem (10)–(13) is solvable.

LEMMA II. *For the functions $W_{j\ell}$ for even $j + \ell$ we have the representations*

$$W_{0l} = -z^{\frac{l}{2}}\delta(t - z) + \ldots, \tag{14}$$

$$W_{jl} = S_{jl}(z)\,\varepsilon(t - z) + \ldots \quad (j = 1,\ 2,\ \ldots k), \tag{15}$$

where $S_{j\ell}$ are smooth functions of z which for $\ell = 0$ or 1 satisfy the expression

$$S'_{j\sigma_j} = -\frac{1}{2}\,q_j(z)\cdot z^{\frac{j+\sigma_j}{2}} \quad (j = 1,\ 2,\ \ldots k). \tag{16}$$

The dots in (14) and (15) denote terms having a weaker singularity on the characteristic than the terms which are written out.

LEMMA III. *For $W_{0\ell}$ above the characteristic in the neighborhood of the coordinate origin, the representation (for even $\ell = 2s > 0$)*

$$W_{0l}(\zeta,\ 2z - \zeta) = -\frac{1}{2\,(s-1)!}\,\frac{1}{z}\,\frac{d^s}{dz^s}\,[z^{s+1}(z - \zeta)^{s-1}] + O(z^s) \tag{17}$$

is valid for $0 < \zeta < z$.

LEMMA IV. *The functions W_{js} in the neighborhood of the coordinate origin satisfy the estimate*

$$W_{js}(z,\ t) = O\left((z + t)^{\frac{j+s}{2}+1}\right) \quad (j = 1,\ 2,\ \ldots k). \tag{18}$$

In particular, Lemma IV implies that conditions (5) on $b_k(t)$ are necessary conditions for the problem to be solvable. The proof of (14)–(18) is given in the Appendix. Here we will consider some corollaries of them.

Using (15) and (16), we may write

$$-\frac{1}{2}\,q_k(z)\,z^{\frac{k+\sigma_k}{2}} = \frac{d}{dz}\,[W_{k\sigma_k}(z,\ z)]. \tag{19}$$

Here and henceforth (as in similar situations in [5–7]) we understand $W_{k\sigma_k}(z, z)$ to be the limit of $W_{k\sigma_k}(z, t)$ as the point (z, t) tends to the characteristic from the direction in which this limit is nonzero (as $t \to z + 0$ in our case).

On the other hand, we use the fact that $W_{k\sigma_k}$ satisfies (10) and the "initial data" (12)–(13).

Let $G(z, \zeta, t)$ be the solution of the direct problem

$$G_{tt} = G_{zz} + q_0(z)G + \delta(t)\delta(z-\zeta)$$
$$G|_{z<\zeta} \equiv 0.$$

and let $G_0(z, t)$ be the solution of the problem (also direct)

$$G_{0tt} = G_{0zz} + q_0(z)G_0,$$
$$G_0|_{z=0} = \delta(t), \quad G_{0z}|_{z=0} = 0.$$

Then

$$W_{k\sigma_k}(z, t) = \int\int G(z, \zeta, t-\tau)q_k(\zeta)W_{0k+\sigma_k}(\zeta, \tau)\,d\zeta d\tau$$
$$+\int\int G(z, \zeta, t-\tau)\Psi_k(\zeta, \tau)\,d\zeta d\tau + \int G_0(z, t-\tau)b_k(\tau)\,d\tau: \qquad (20)$$

Substituting $W_{k\sigma_k}$ from (20) into (19), we obtain the integral equation

$$-\frac{1}{2}q_k(z)z^{\frac{k+\sigma_k}{2}} = \int\left[\frac{d}{dz}\int_{-\infty}^{\infty} G(z, \zeta, z-\tau)W_{0k+\sigma_k}(\zeta, \tau)\,d\tau\right]q_k(\zeta)\,d\zeta + \omega_k(z), \qquad (21)$$

where

$$\omega_k(z) \equiv \frac{d}{dz}\int\int G(z, \zeta, z-\tau)\Psi_k(\zeta, \tau)\,d\zeta d\tau + \frac{d}{dz}\int G_0(z, z-\tau)b_k(\tau)\,d\tau. \qquad (22)$$

Equation (21) is a singular Volterra equation (the coefficient for the unknown function $q_k(z)$ vanishes); the kernel $L_k(z, \zeta) = (d/dz)\int G(z, \zeta, z-\tau)W_{0k+\sigma_k}(\zeta, \tau)d\tau$ is obviously a Volterra kernel, since $G(z, \zeta, t) \equiv 0$ for $\zeta > z$. It may be readily shown that $G(z, \zeta, t) = \frac{1}{2}\varepsilon(z-\zeta+t)\varepsilon(z-\zeta-t)\theta(z, \zeta, t)$, where $\theta(z, \zeta, t)$ is a smooth function equal to 1 for $z-\zeta = \pm t$. Therefore

$$L_k(z, \zeta) = \frac{d}{dz}\int_{\zeta}^{2z-\zeta} G(z, \zeta, z-\tau)W_{0k+\sigma_k}(\zeta, \tau)\,d\tau$$
$$= 2G(z, \zeta, \zeta-z)W_{0k+\sigma_k}(\zeta, 2z-\zeta) + \int_{\zeta}^{2z-\zeta}\frac{d}{dz}G(z, \zeta, z-\tau)W_{0k+\sigma_k}(\zeta, \tau)\,d\tau,$$

and $dG(z, \zeta, z-\tau)/dz$ is a classical bounded function for $\zeta < \tau < 2z-\zeta$. Allowing for (17), therefore, we have

$$L_k(z, \zeta) = M_k(z, \zeta) + N_k(z, \zeta),$$

where

$$M_k(z, \zeta) = \frac{1}{2} \frac{1}{(s_k - 1)!} \frac{1}{z} \frac{d^{s_k}}{dz^{s_k}} [z^{s_k+1} (z - \zeta)^{s_k - 1}], \tag{23}$$

and

$$N_k(z, \zeta) = 0(z^{s_k}), \quad s_k = \frac{k + \sigma_k}{2}.$$

As a result, equation (21) becomes

$$0 = q_k z^{s_k} + \frac{1}{(s_k - 1)!} \frac{1}{z} \int_0^z \frac{d^{s_k}}{dz^{s_k}} [z^{s_k+1} (z - \zeta)^{s_k-1}] q_k(\zeta)\, d\zeta$$

$$+ 2 \int_0^z N_k(z, \zeta) q_k(\zeta)\, d\zeta + 2\omega_k(z).$$

The first two terms of the last equation may obviously be rewritten as

$$\frac{1}{(s_k - 1)!} \frac{1}{z} \frac{d^{s_k}}{dz^{s_k}} z^{s_k+1} \int_0^z (z - \zeta)^{s_k-1} q_k(\zeta)\, d\zeta,$$

whence, after elementary operations, we obtain

$$q_k(z) = 2 \frac{d^{s_k}}{dz^{s_k}} \frac{1}{z^{s_k+1}} \int_0^z \frac{(z - z_1)^{s_k-1}}{(s_k - 1)!} z_1 dz_1 \int_0^{z_1} N_k(z_1, \zeta) q_k(\zeta)\, d\zeta + 2\tilde{\omega}_k(z), \tag{24}$$

where

$$\tilde{\omega}_k(z) = \frac{d^{s_k}}{dz^{s_k}} \frac{1}{z^{s_k+1}} \int_0^z \frac{(z - z_1)^{s_k-1}}{(s_k - 1)!} z_1 \omega_k(z_1)\, dz_1.$$

Equation (24) has the form

$$q_k(z) - R_k q_k + 2\tilde{\omega}_k.$$

Let us study the kernel $R_k(z, \zeta)$ of this equation. Obviously

$$R_k(z, \zeta) = 2 \frac{d^{s_k}}{dz^{s_k}} \frac{1}{z^{s_k+1}} \int_\zeta^z \frac{(z - z_1)^{s_k-1}}{(s_k - 1)!} z_1 N_k(z_1, \zeta)\, dz_1. \tag{25}$$

Let us estimate $R_k(z, \zeta)$. Differentiating for the kernel in (25), we see that $R_k(z, \zeta)$ is a finite sum of terms of the form

$$A_{kj} \frac{1}{z^{s_k+1+j}} \int_\zeta^z \frac{(z - z_1)^{j-1}}{(j - 1)!} z_1 N_k(z_1, \zeta)\, dz_1 \quad (j = 1, \ldots s_k),$$

each of which does not exceed

$$CA_{kj}\frac{1}{z^{s_k+1+j}}\int_{\zeta}^{z}\frac{(z-z_1)^{j-1}}{(j-1)!}z_1^{s_k+1}dz_1$$

$$\leqslant \frac{CA_{kj}}{(j-1)!}\frac{1}{z^{s_k+1+j}}\int_0^z (z-z_1)^{j-1}z_1^{s_k+1}dz_1=\frac{CA_{kj}}{(j-1)!}B(j,\ s_k+2)$$

in absolute value, and of the term $2N_k(z,\zeta)/z^{s_k}$, which is also bounded. Thus (24) is an integral equation of the second kind of Volterra type with a bounded (and even continuous) kernel. Our study of this equation will be complete if we can show that the free term $2\tilde{\omega}_k(z)$ is a continuous function of z. It is clear that this is the case if $\omega_k(z)$ is a continuous function of z which satisfies $|\omega_k(z)| \leqslant Cz^{s_k}$ in the neighborhood of the point $z = 0$. Consider expression (22) for $\omega_k(z)$. We note that $G(z,\zeta,z-\tau)$ experiences a jump equal to 1/2 on the lines $z-\zeta = \pm(z-\tau)$. Therefore

$$\frac{dG(z,\ \zeta,\ z-\tau)}{dz}=-\delta(2z-\zeta-\tau)\,\varepsilon(z-\zeta)+\ldots,$$

where the dots denote a classical (bounded) contribution which is equal to zero outside the triangle $2z-\zeta \geqslant \tau \geqslant \zeta \geqslant 0$, and since in view of Lemma IV we have $\Psi_k = O((t+z)^{(k+\sigma_k)/2+1})$, the first term on the right side of (22) is $O(z^{(k+\sigma_k)/2+2})$.

The function $G_0(z, z-\tau)$ has a δ-singularity at the points $\tau = 0$ and $\tau = 2z$. Therefore

$$\frac{d}{dz}\int G_0(z,\ z-\tau)\,b_k(\tau)\,d\tau=\frac{d}{dz}\frac{1}{2}b_k(2z)$$

$$+\frac{d}{dz}\int_0^{2z}G(z,\ z-\tau)\,b_k(\tau)\,d\tau=b_k'(2z)+\frac{d}{dz}\int_0^{2z}G_0(z,\ z-\tau)\,b_k(\tau)\,d\tau,$$

where the integral is understood to be the integral taken from the classical function (the δ-singularity is isolated in the first term). Conditions (6) on $b_k(t)$ imply that the second term in (22), and hence $w_k(z)$, is $O(z^{s_k})$, QED.

Appendix

In this Appendix we will prove Lemmas II–IV. These lemmas establish the properties of the functions W_{kl} for even $k+l$,[1] where the W_{kl} are regarded as solutions of the direct problems

$$W_{kltt}=W_{klzz}+q_0W_{kl}+\frac{l(l-1)}{k+l-1}W_{kl-2t}+\sum_{i=1}^{k-1}q_iW_{k-i\,l+i}+q_kW_{0\,l+k},\qquad (1^0)$$

$$W_{klz}|_{z=0}=0\quad \text{for}\quad k+l>0\qquad (2^0)$$

or

$$W_{00z}|_{z=0}=\delta'(t)\quad \text{for}\quad k=l=0.\qquad (2^0\text{a})$$

[1] Henceforth we will encounter W_{kl} only for even $k+l$. Therefore this fact will not be stipulated each time.

As always, the conditions

$$W_{kl}\,|_{t<0}\equiv 0 \tag{3^0}$$

are understood. Let $k = \ell = 0$. Then for W_{00} we obviously have

$$W_{00}=S_{00}(z)\,\delta(t-z)+\Phi_{00}(z,\ t). \tag{4^0}$$

Here $\Phi_{00}(z, t)$ denotes a function which has a lesser singularity as compared to the term written out. We find the coefficient S_{00} by substituting (4^0) into (1^0) and equating the coefficients for the largest of the singular terms (in this case for $\delta'(t-z)$). Thus we find a differential equation for $S_{00}(z)$, namely $2S'_{00} = 0$, and, allowing for condition (2^0a), $-S_{00}(z) = -1$. The fact that the singularity of $\Phi_{00}(z, t)$ is indeed less than that of the first term in (4^0) (more precisely, that $\Phi_{00}(z,t)$ is a classical bounded function) may be readily proved by using the integral equations. For the moment we will not do this. We will formally calculate the singularity of the functions $W_{0\ell}$. Let $W_{0s} = -z^{s/2}\delta(t-z) + \Phi_{0s}$ for $s < \ell$. Then for the $W_{0\ell}$ we have the problem

$$W_{0ltt}=W_{0lzz}+q_0W_{0l}-lz^{l/2-1}\delta'(t-z)+l\Phi_{0l-2t},$$
$$W_{0lz}\,|_{z=0}=0.$$

Hence

$$W_{0l}=S_{0l}(z)\,\delta(t-z)+\Phi_{0l}. \tag{5^0}$$

Substituting this expression into the equation for $W_{0\ell}$, we find (retaining only the terms containing $\delta''(t-z)$ and $\delta'(t-z)$) that

$$S_{0l}\delta''=S_{0l}\delta''-2S'_{0l}\delta'-lz^{l/2-1}\delta',$$

whence

$$S'_{0l}=-\frac{l}{2}\,z^{l/2-1}$$

and (allowing for the boundary term)

$$S_{0l}=-z^{l/2}.$$

Thus formula (14) has been proved correct to the estimate of the remainder term. To prove (15) we will assume that it has been proved for W_{js} for all $0 < j < k$, $s \geqslant 0$ and for W_{ks} for $s < \ell$. Then, substituting representations (14) and (15) into (1^0) for those terms for which it is proved, and continuing all functions in (1^0) evenly onto negative z, we obtain

$$W_{kltt}=W_{klzz}+q_0W_{kl}+\frac{l\,(l-1)}{k+l-1}S_{kl-2}\delta(t-|z|)$$
$$-q_k|z|^{l+k/2}\delta(t-|z|)+\frac{l\,(l-1)}{k+l-1}\Phi_{kl-2t}+q_k\Phi_{0l+k}+\sum_{i=1}^{k-1}q_iW_{k-il+i}, \tag{6^0}$$

or for $t > |z|$

$$W_{kl} = \frac{1}{2} \iint_{\square zt} W_{kl}(\zeta, \tau)\, q_0(\zeta)\, d\zeta d\tau + \frac{l(l-1)}{2(k+l-1)} \left[\int_0^{(t+z)/2} S_{lk-2}(\zeta)\, d\zeta + \int_0^{(t-z)/2} S_{kl-2}(\zeta)\, d\zeta \right]$$
$$- \frac{1}{2} \int_0^{(t+z)/2} q_k(\zeta)\, \zeta^{l+k/2} d\zeta - \frac{1}{2} \int_0^{(t-z)/2} q_k(\zeta)\, \zeta^{l+k/2} d\zeta + \frac{1}{2} F_{kl}(z, t), \qquad (7^0)$$

where $\square_{zt}$ is the rectangle $-0 \leqslant \zeta + \tau < z + t$, $-0 \leqslant \tau - \zeta < t - z$; $F_{kl}(z,t)$ is the integral of the last three terms on the right side of (6^0) with respect to $\square_{zt}$. Assertion (15) for W_{kl} is obvious from integral equation (7^0). We note that if W_{js} for $0 < j < k$ and W_{ks} for $s < l$ satisfy estimate (18) and $\Phi_{0s} = O((t+z)^{s/2-1})$, then $W_{kl} = O((t+z)^{(k+l)/2+1})$ (all estimates are for $z > 0$). Setting $l = \sigma_k$, $k = 0$ and $t = z + 0$ in (7^0), we obtain

$$W_{k\sigma_k}|_{t=z+0} = S_{k\sigma_k}(z) = -\frac{1}{2} \int_0^z q_k(\zeta)\, \zeta^{(\sigma_k+k)/2} d\zeta,$$

whence formula (16) is obvious. Thus all the assertions of Lemmas II and IV will be proved if we prove the estimate for Φ_{0s}. But this is a trivial corollary of Lemma III, which we will now proceed to prove. Substituting the representation $W_{0l} = -z^{l/2}\delta(t-z) + \Phi_{0l}$ into problem (1^0) – (3^0), we obtain for Φ_{0l} the problem

$$\Phi_{0ltt} = \Phi_{0lzz} + q_0\Phi_{0l} + l\Phi_{0l-2t} - \frac{l(l-2)}{4} |z|^{l/2-2}\delta(t-|z|) - q_0|z|^{l/2}\delta(t-|z|). \qquad (8^0)$$

Here we have avoided the condition $\Phi_{0lz}|_{z=0} = 0$ by continuing all functions evenly onto negative z. For $l = 0$ and 2 the change to problem (8^0) requires greater care. Let $l = 0$. Continuing the solution W_{00} of problem (1^0), $(2^0\mathrm{a})$, (3^0) evenly onto negative z, we readily obtain that W_{00} is the solution of the problem

$$W_{00tt} = W_{00zz} + q_0 W_{00} - 2\delta'(t)\,\delta(z).$$

If we allow for the fact that the function $-\delta(t-|z|)$ is the solution of the same problem with $q_0 \equiv 0$, then $\Phi_{00} = W_{00} + \delta(t-|z|)$ is the solution of the problem

$$\Phi_{00tt} = \Phi_{00zz} + q_0\Phi_{00} - q_0(z)\,\delta(t-|z|),$$

which coincides with (8^0) if we allow for the fact that the terms in the l which appears as a factor are equal to 0.

For $\ell = 2$, W_{02} is the solution of the problem

$$W_{02tt} = W_{02zz} + q_0 W_{02} + 2W_{00t}$$

(we assume that the even continuation with respect to z has already been made). Replacing $W_{0\ell}$ for $\ell = 0$ and 2 by the representation $W_{0\ell} = -|z|^{\ell/2}\delta(t-|z|) + \Phi_{0\ell}$, and allowing for the fact that

$$\left(\frac{\partial^2}{\partial t^2} - \frac{\partial^2}{\partial z^2}\right)\left[-|z|\delta(t-|z|)\right] = 2\delta(z)\delta(t) - 2\delta'(t-|z|),$$

we obtain the problem for Φ_{02} :

$$\Phi_{02tt} = \Phi_{02zz} + q_0\Phi_{02} + 2\Phi_{00t} - 2\delta(z)\delta(t) - |z|q_0(z)\delta(t-|z|),$$

which coincides with (8^0) if in the fourth term on the right side of (8^0) we understand $\ell(\ell-2)|z|^{\ell/2-2}/4$ to be $\partial^2|z|^{\ell/2}/\partial z^2$ (obviously, the last expression for $\ell - 2$ is equal to $2\delta(z)$).

For brevity we set $\Phi_{0,2s} \equiv \Phi_s$. For Φ_s we must consider the problem

$$\Phi_{stt} = \Phi_{szz} + q_0\Phi_s + 2s\Phi_{s-1t} - (|z|^s)''\delta(t-|z|) - q_0(z)|z|^s\delta(t-|z|). \quad (9^0)$$

Problem (9^0) leads to the integral equation (for $t > |z|$)

$$\Phi_s = \frac{1}{2}\iint\limits_{\square zt} q_0\Phi_s d\zeta d\tau + s\iint\limits_{\square zt} \Phi_{s-1\tau} d\zeta d\tau$$
$$- \frac{1}{2}\iint\limits_{\square zt} (|\zeta|^s)''\delta(\tau-|\zeta|)\,d\zeta d\tau - \frac{1}{2}\iint\limits_{\square zt} q_0(\zeta)|\zeta|^s\delta(\tau-|\zeta|)\,d\zeta d\tau,$$

or

$$\Phi_s = \frac{1}{2}\iint\limits_{\square zt} q_0\Phi_s d\zeta d\tau + s\int_z^{(z+t)/2} \Phi_{s-1}(\zeta, z+t-\zeta)\,d\zeta$$
$$+ s\int_{(z-t)/2}^{z} \Phi_{s-1}(\zeta, t-z+\zeta)\,d\zeta - M_s - \frac{1}{2}\int_0^{(t+z)/2} q(\zeta)\zeta^s d\zeta - \frac{1}{2}\int_0^{(t-z)/2} q(\zeta)\zeta^s d\zeta, \quad (10^0)$$

where

$$M_s = \begin{cases} 0 & \text{for } s = 0, \\ \frac{s}{2}\left[\left(\frac{t+z}{2}\right)^{s-1} + \left(\frac{t-z}{2}\right)^{s-1}\right] & \text{for } s \geqslant 1. \end{cases}$$

It is obvious from (10^0) that $\Phi_0 = O(t+z)$ (henceforth we will assume that $z > 0$).

Let $s = 1$. Then

$$\Phi_1=\frac{1}{2}\iint_{\square zt} q_0\Phi_1 d\zeta d\tau-\int_z^{(z+t)/2}\Phi_0(\zeta,\ z+t-\zeta)\,d\zeta$$

$$-\int_{(z-t)/2}^{z}\Phi_0(\zeta,\ t-z+\zeta)\,d\zeta-1-\frac{1}{2}\int_0^{(z+t)/2} q(\zeta)\,\zeta d\zeta-\frac{1}{2}\int_0^{(z-t)/2} q(\zeta)\,\zeta d\zeta.$$

Consequently

$$\Phi_1=-1+O((t+z)^2), \tag{11^0}$$

and (17) is proved for $s=1$ (we have even proved a slightly stronger assertion). We will assume that we have proved the expression

$$\Phi_{s-1}=\frac{1}{2}P_{s-2}(z,\ t)+O((t+z)^s),$$

where $P_{s-2}(z,t)$ is a homogeneous polynomial of degree $s-2$. Equation (10^0) then implies that

$$\Phi_s=\frac{1}{2}\iint q_0\Phi_s d\zeta d\tau-s\int_z^{(z+t)/2}P_{s-2}(\zeta,\ z+t-\zeta)\,d\zeta$$

$$-s\int_{(z-t)/2}^{z}P_{s-2}(\zeta,\ t-z+\zeta)\,d\zeta-\frac{s}{2}\left[\left(\frac{t+z}{2}\right)^{s-1}+\left(\frac{t-z}{2}\right)^{s-1}\right]+O((t+z)^{s+1}).$$

Thus the representation

$$\Phi_s=\frac{1}{2}P_{s-1}(z,\ t)+O((t+z)^{s+1})$$

has been proved; we have obtained a recurrence relation for the polynomials:

$$P_{s-1}(z,\ t)=s\int_z^{(z+t)/2}P_{s-2}(\xi,\ z+t-\xi)\,d\xi$$

$$+s\int_{(z-t)/2}^{z}P_{s-2}(\xi,\ t-z+\xi)\,d\xi-s\left[\left(\frac{t+z}{2}\right)^{s-1}+\left(\frac{t-z}{2}\right)^{s-1}\right].$$

This formula may be conveniently written as follows:

$$P_{s-1}(\zeta,\ 2z-\zeta)=s\int_\zeta^{z}P_{s-2}(\xi,\ 2z-\xi)\,d\xi$$

$$+s\int_{\zeta-z}^{\zeta}P_{s-2}(\xi,\ 2z-2\zeta+\xi)\,d\xi-s\,(z^{s-1}-(z-\zeta)^{s-1})$$

or, replacing ζ by the variable $z - \zeta = y$ and setting $P_{s-1}(\zeta, 2z - \zeta) = P_{s-1}(z - y, z + y) = Q_{s-1}(z, y)$,

$$Q_{s-1}(z, y) = s\int_{z-y}^{z} Q_{s-2}(z, z-\xi)\,d\xi + s\int_{-y}^{z-y} Q_{s-2}(\xi + y, y)\,d\xi - s(z^{s-1} + y^{s-1})$$

$$= s\int_0^y Q_{s-2}(z, \xi)\,d\xi + s\int_0^z Q_{s-2}(\xi, y)\,d\xi - s(z^{s-1} + y^{s-1}). \qquad (12^0)$$

Let us check to see that the formula

$$Q_{k-1}(z, y) = -\frac{k}{2\pi i}\oint \frac{(1+\lambda)^{k+1}(y + z\lambda)^{k-1}}{\lambda^{k+1}}\,d\lambda, \qquad (13^0)$$

is valid, where the integral is taken over a contour enclosing the point $\lambda = 0$ (on the plane of the complex variable λ).

Indeed, for $s = 1$ we obtain

$$Q_0(z, y) = -\frac{1}{2\pi i}\oint \frac{(1+\lambda)^2}{\lambda^2}\,d\lambda = -2$$

in accordance with (11^0).

We will show that if (13_0) holds for $k = s - 1$, it is valid for $k = s$. In accordance with (13^0),

$$s\int_0^y Q_{s-2}(z, \xi)\,d\xi - sz^{s-1} = -\frac{s(s-1)}{2\pi i}\int_0^y d\xi \oint \frac{(1+\lambda)^s(\xi + z\lambda)^{s-2}}{\lambda^s}\,d\lambda - sz^{s-1}$$

$$= -\frac{s}{2\pi i}\oint \frac{(1+\lambda)^s(\xi + z\lambda)^{s-1}}{\lambda^s}\,d\lambda\Big|_{\xi=0}^{\xi=y} - sz^{s-1}$$

$$= -\frac{s}{2\pi i}\oint \frac{(1+\lambda)^s}{\lambda^s}(y + z\lambda)^{s-1}\,d\lambda + \frac{s}{2\pi i}\oint \frac{(1+\lambda)^s}{\lambda^s}(z\lambda)^{s-1}\,d\lambda - sz^{s-1}$$

$$= -\frac{s}{2\pi i}\oint \frac{(1+\lambda)^s}{\lambda^s}(y + z\lambda)^{s-1}\,d\lambda$$

and

$$s\int_0^z Q_{s-2}(\xi, y)\,d\xi - sy^{s-1} = -\frac{s(s-1)}{2\pi i}\int_0^z d\xi \oint \frac{(1+\lambda)^s(y + \xi\lambda)^{s-2}}{\lambda^s}\,d\lambda - sy^{s-1}$$

$$= -\frac{s}{2\pi i}\oint \frac{(1+\lambda)^s(y + z\lambda)^{s-1}}{\lambda^{s+1}}\,d\lambda + \frac{s}{2\pi i}\oint \frac{(1+\lambda)^s y^{s-1}}{\lambda^{s+1}}\,d\lambda - sy^{s-1}$$

$$= -\frac{s}{2\pi i}\oint \frac{(1+\lambda)^s(y + z\lambda)^{s-1}}{\lambda^{s+1}}\,d\lambda.$$

Adding the last two expressions and using recurrence relation (12^0), we find

$$Q_{s-1}(z,\ y)=s\int_0^y Q_{s-2}(z,\ \xi)\,d\xi+s\int_0^z Q_{s-2}(\xi,\ y)\,d\xi-sz^{s-1}-sy^{s-1}$$

$$=-\frac{s}{2\pi i}\oint(1+\lambda)^s(y+z\lambda)^{s-1}\left(\frac{1}{\lambda^{s+1}}+\frac{1}{\lambda^s}\right)d\lambda$$

$$=-\frac{s}{2\pi i}\oint\frac{(1+\lambda)^{s+1}(y+z\lambda)^{s-1}}{\lambda^{s+1}}\,d\lambda,$$

which proves (13^0).

Thus we have found an expression for $Q_{s-1}(z,y)$. We transform it by changing to the variables z and ζ:

$$Q_{s-1}(z,\ z-\zeta)=-\frac{s}{2\pi i}\oint\frac{(1+\lambda)^{s+1}(z-\zeta+\lambda z)^{s-1}}{\lambda^{s+1}}\,d\lambda$$

$$=-\frac{s}{2\pi iz}\oint\frac{\left(1+\frac{\lambda_1}{z}\right)^{s+1}}{\lambda_1^{s+1}}\,z^{s+1}(z-\zeta+\lambda_1)^{s-1}\,d\lambda_1$$

$$=-\frac{s}{2\pi iz}\oint\frac{(z+\lambda_1)^{s+1}}{\lambda_1^{s+1}}(z-\zeta+\lambda_1)^{s-1}\,d\lambda_1$$

$$=-\frac{s}{s!\,z}\frac{d^s}{d\lambda_1^s}(z+\lambda_1)^{s+1}(z-\zeta+\lambda_1)^{s-1}\Big|_{\lambda_1=0}=-\frac{1}{(s-1)!\,z}\frac{d^s}{dz^s}[z^{s+1}(z-\zeta)^{s-1}].$$

Formula (17) is thus proved.

BIBLIOGRAPHY

1. M.G. Kreĭn, *On a method of effective solution of an inverse boundary problem*, Dokl. Akad. Nauk SSSR **94** (1954), 987–990. (Russian) MR **16**, 38.
2. I.M. Gel'fand and B.M. Levitan, *On the determination of a differential equation from its spectral function*, Izv. Akad. Nauk SSSR Ser. Mat. **15** (1951), 309–360; English transl., Amer. Math. Soc. Transl. (2) **1** (1955), 253–304. MR **13**, 558; **17**, 489.
3. L.D. Faddeev, *The inverse problem in the quantum theory of scattering*, Uspehi Mat. Nauk **15** (1959), no. 4 (88), 57–119; English transl., J. Mathematical Phys. **4** (1963), 72–104. MR **22** #1344; **26** #7328.
4. A.S. Blagoveščenskiĭ, *An inverse problem in the theory of propagation of seismic waves*, Problemy Mat. Fiz., no. 1, Izdat. Leningrad. Univ., Leningrad, 1966, pp. 68–81; English transl., Topics in Math. Phys., no. 1, Consultants Bureau, New York, 1967, pp. 55–67.
5. ———, *The inverse boundary value problem in the theory of wave propagation in an anisotropic medium*, Trudy Mat. Inst. Steklov. **115** (1971), 39–56 = Proc. Steklov Inst. Math. **115** (1971), 41–62.
6. A.A. Buzdin, *An inverse problem in a medium with absorption*, Trudy Mat. Inst. Steklov. **115** (1971), 70–77 = Proc. Steklov Inst. Math. **115** (1971), 77–85.
7. A.S. Blagoveščenskiĭ, *A local method of solving the nonstationary inverse problem for a nonuniform string*, Trudy Mat. Inst. Steklov **115** (1971), 28–38 = Proc. Steklov Inst. Math. **115** (1971), 28–40.

Translated by:
M. Lopez-Morillas

ON AN INVERSE PROBLEM IN A MEDIUM WITH ABSORPTION

A. A. BUZDIN

UDC 517 944/947

Abstract. The article considers the problem of finding the coefficients of the equation

$$Lu = u_{tt} + \sigma(z)\, u_t - c^2 \Delta u,$$

if the function u satisfies the conditions

$$u\,|_{t<0} \equiv 0, \quad u_z\,|_{z=0} = \delta(t)\,\delta(x, y)$$

and, moreover, certain characteristics of the function $u|_{z=0} = F(t, x, y)$ are known. The problem can be reduced to certain integro-differential equations which may then be studied further.

Figures: 1. Bibliography: 2 items.

§1. Derivation of the integro-differential equations

In this paper we will investigate the following inverse problems: for the equation

$$\frac{\partial^2}{\partial t^2} U(t, x, y, z) + \sigma(z) \frac{\partial}{\partial t} U(t, x, y, z) = C^2(z)\, \Delta U(t, x, y, z) \tag{1}$$

with initial conditions

$$U\,|_{t<0} \equiv 0 \tag{2}$$

and the boundary condition

$$U_z\,|_{z=0} = \delta(t)\,\delta(x, y) \tag{3}$$

we are to determine the functions $\sigma(z)$ and $C(z)$. We assume that certain characteristics of the function $F(t, x, y)$ are given:

$$U\,|_{z=0} = F(t, x, y). \tag{4}$$

In physical problems the coefficient $\sigma(z)$, subject to the condition that $\sigma(z) > 0$, is responsible for wave attenuation. We will not assume that this condition holds; however, we do note that because the equation is hyperbolic, the function $U(x, y, z, t)$ is finite for any fixed t, and therefore any of its moments have a meaning. To solve our problem we will require the following two moments:

$$V(z, t) = \int\int U(x, y, z, t)\, dx\,dy,$$

$$W(z, t) = \int\int x^2 U(x, y, z, t)\, dx\,dy.$$

AMS 1970 *subject classifications*. Primary 45K05; Secondary 73D99.

The first moment $\iint xU(x,y,z,t)dxdy$ will be ignored, since it is necessarily equal to zero. We will assume that instead of $F(t,x,y)$ we are given the values of the functions V and W for $z = 0$:

$$V(z,t)|_{z=0} = f(t), \tag{5}$$

$$W(z,t)|_{z=0} = g(t). \tag{6}$$

It is easy to obtain a system of differential equations for $V(z,t)$ and $W(z,t)$:

$$\left.\begin{aligned} \frac{\partial^2 V}{\partial t^2} + \sigma(z)\frac{\partial V}{\partial t} &= C^2(z)\frac{\partial^2 V}{\partial z^2}, \\ \frac{\partial^2 W}{\partial t^2} + \sigma(z)\frac{\partial W}{\partial t} &= C^2(z)\frac{\partial^2 W}{\partial z^2} + 2C^2(z)V. \end{aligned}\right\} \tag{7}$$

The initial and boundary conditions for V and W have the following form:

$$\begin{gathered} V|_{z=0} = f(t), \quad W|_{z=0} = g(t), \\ V_z|_{z=0} = \delta(t), \quad W_z|_{z=0} = = 0, \\ V|_{t<0} \equiv 0, \quad W|_{t<0} \equiv 0. \end{gathered} \tag{8}$$

We now introduce the new variable

$$y = \int_0^z \frac{dz}{C(z)}.^{1)} \tag{9}$$

After some operations, problem (7)–(8) becomes the following:

$$V_{tt} + \sigma(y)V_t = V_{yy} - K(y)V_y, \tag{10}$$

$$W_{tt} + \sigma(y)W_t = W_{yy} - K(y)W_y + 2C^2(y)V, \tag{11}$$

$$V|_{y=0} = f(t), \quad V_y|_{y=0} = C(0)\delta(t), \tag{12}$$

$$W|_{y=0} = g(t), \quad W_y|_{y=0} = 0, \tag{13}$$

$$V|_{t<0} \equiv 0, \quad W|_{t<0} \equiv 0, \tag{14}$$

Here $K(y) = (\ln C(y))'_y$. We will solve problem (10)–(14) under the assumption that $f(t) \in C_2$, $g(t) \in C_3$, and we will show that among the functions $\sigma(y)$, $K(y) \in C_1$, $V(y,t) \in C_{2,2}$, $W(y,t) \in C_{2,3}$ there exists a unique solution of the problem over a finite depth $y < y_0$.

Here $C_{m,m+n}$ $(m \neq 0)$ is the class of functions which are m times continuously differentiable, which are defined for $t > y$, and which have, moreover, continuous derivatives of the form $\partial^{m+i}U/\partial y\partial t^{m+i-1}$ $(i = 1,2,\dots,n)$ and $\partial^{m+i}U/\partial t^{m+i}$ $(i = 1,2,\dots,n)$. C_{0n} is the class of continuous functions defined for $t > y$ which have continuous derivatives in t up to order n inclusive.

---

1) Of course, here and henceforth y does not coincide with the independent variable y of equation (1).

We will now solve the problem for equations (10) and (11) with the Cauchy data (12), (13) for $y = 0$, assuming the functions $\sigma(y)V_t$, $K(y)V_y$, $\sigma(y)W_t$, $K(y)W_y$ and $C^2(y)V$ to be known:

$$V(y, t) = \frac{1}{2} \iint_{\Delta y, t} [\sigma(\eta) V_\tau(\eta, \tau) + K(\eta) V_\eta(\eta, \tau)]\, d\eta d\tau + \frac{1}{2}[f(t+y) + f(t-y)], \qquad (15)$$

$$W(y, t) = \frac{1}{2} \iint_{\Delta y, t} [\sigma(\eta) W_\tau(\eta, \tau) + K(\eta) W_\eta(\eta, \tau)]\, d\eta d\tau - \iint_{\Delta y, t} C^2(\eta) V(\eta, \tau)\, d\eta d\tau + \frac{1}{2}[g(t+y) + g(t-y)], \qquad (16)$$

where $\Delta y, t$ is the triangle formed by the characteristics of (10) and (11) which pass through the point y, t and by the axis $y = 0$.

The following problem will be called the direct problem of (10)–(14): to find the functions V and W for equations (10) and (11) if we are given $\sigma(y)$ and $K(y)$ as well as the initial conditions (14) and the boundary conditions

$$V_y\big|_{y=0} = C(0)\delta(t), \quad W_y\big|_{y=0} = 0.$$

The properties of the solution of the direct problem imply that if the coefficients $\sigma(y)$ and $K(y)$ belong to C, then $f(t) \in C_2$, $g(t) \in C_3$, $V(y,t) \in C_{2,2}$, $W(y,t) \in C|_{2,3}$, where

$$V(y, t) = A_0(y)\varepsilon(t-y) + A_1(y)(t-y)_+ + V_1(y, t), \qquad (17)$$

$$W(y, t) = B_0(y)\varepsilon(t-y) + B_1(y)(t-y)_+ + B_2(y)(t-y)_+^2 + W_1(y, t). \qquad (18)$$

Here A_i, $B_i \in C_2$, while V_1 and W_1 have a singularity on the characteristic which is less than the singularity of the last term written out in expansions (17) and (18). It is in this form that we will seek the solution of our system of equations.

Substituting (17) and (18) into (10) and (11) and equating the coefficients for $\delta(t-y)$, $\varepsilon(t-y)$ and $(t-y)_+$, we obtain differential equations for the coefficients $A_0(y)$, $A_1(y)$, $B_0(y)$, $B_1(y)$ and $B_2(y)$.

It is easy to see that $B_0 = 0$. This is a consequence of the equation

$$B_0' = -\frac{\sigma(y) - K(y)}{2} B_0$$

and of boundary condition (13).

If we introduce instead of $\sigma(y)$ and $K(y)$ the functions

$$\begin{aligned} X(y) &= \sigma(y) - K(y), \\ Z(y) &= \sigma(y) + K(y), \end{aligned} \qquad (19)$$

then the equations for the coefficients have the following form:

$$A_0' = -\frac{X}{2} A_0, \tag{20}$$

$$A_1' = -\frac{X}{2} A_1 + \left(\frac{XZ}{8} - \frac{X'}{4}\right) A_0, \tag{21}$$

$$B_1' = -\frac{X}{2} B_1 + C^2(y) A_0, \tag{22}$$

$$B_2' = -\frac{X}{2} B_2 + \frac{B_1''}{4} + \frac{X-Z}{8} B_1' + \frac{1}{2} C^2(y) A_1. \tag{23}$$

We introduce the new unknown functions $C_1(y)$, $D_1(y)$ and $D_2(y)$ by the expressions

$$A_1(y) = C_1(y) A_0(y),$$
$$B_1(y) = D_1(y) A_0(y),$$
$$B_2(y) = D_2(y) A_0(y).$$

We obtain the following equations for them:

$$C_1'(y) = \frac{XZ}{8} - \frac{X'}{4}, \tag{24}$$

$$D_1'(y) = C^2(y), \tag{25}$$

$$D_2'(y) = \left[-\frac{3}{8} X + \frac{1}{8} Z\right] C^2(y) + \frac{D_1}{8}\left\{\frac{XZ}{2} + X'\right\} + \frac{1}{2} C_1(y) C^2(y). \tag{26}$$

Equations (24)–(26) readily yield expressions for $X(y)$ and $Z(y)$ in terms of A_0, C_1, D_1 and D_2 :

$$X(y) = -\frac{2A_0'}{A_0}, \tag{27}$$

$$Z(y) = \frac{1}{C^2(y)}\{8D_2' - 4C'D_1\} + 3X - 4C_1. \tag{28}$$

The functions C_1 and D_2 are related to the coefficients A_0, A_1, B_1 and B_2 of expansions (17) and (18) for V and W:

$$C_1(y) = \frac{A_1' + \frac{X}{2} A_1}{A_0}, \tag{29}$$

$$D_2(y) = \frac{B_2' + \frac{X}{2} B_2}{A_0}. \tag{30}$$

Integrating (25), we obtain an expression for $D_1(y)$:

$$D_1(y) = \int_0^y C^2(\eta)\, d\eta. \tag{31}$$

In (30) and (31) we will understand $C^2(y)$ and $A_0(y)$ to be the expressions

$$C^2(y) = C^2(0)\exp\left[\int_0^y (Z(\eta) - X(\eta))\,d\eta\right], \tag{32}$$

$$A_0(y) = -C(0)\exp\left[-\frac{1}{2}\int_0^y X(\eta)\,d\eta\right], \tag{33}$$

which are readily obtained from (20) and from the fact that $K(y) = (\ln C(y))'_y$. But, on the other hand,

$$A_0(y) = \lim_{t\to y+0} V(y, t),$$

$$A_1(y) = \lim_{t\to y+0} V_t(y, t),$$

$$B_2(y) = \lim_{t\to y+0} \frac{1}{2} W_{tt}(y, t).$$

Using (15) and (16), we obtain

$$A_0(y) = \frac{1}{2}\iint\limits_{\Delta yy}\left\{\frac{(X+Z)}{2}V_\tau - \frac{(X-Z)}{2}V_\eta\right\}d\tau\,d\eta + \frac{1}{2}\{f(2y) + f(0)\}, \tag{34}$$

$$A_1(y) = \frac{1}{2}\iint\limits_{\Delta yy}\left\{\frac{(X+Z)}{2}V_{\tau\tau} - \frac{(X-Z)}{2}V_{\eta\tau}\right\}d\tau\,d\eta + \frac{1}{2}\{f'(2y) + f'(0)\}, \tag{35}$$

$$B_2(y) = \frac{1}{4}\iint\limits_{\Delta yy}\left\{\frac{(X+Z)}{2}W_{\tau\tau\tau} - \frac{(X-Z)}{2}W_{\eta\tau\tau}\right\}d\tau\,d\eta +$$

$$+ \frac{1}{2}\iint\limits_{\Delta yy}\{C^2(\eta)V_{\tau\tau}\}\,d\tau\,d\eta + \frac{1}{4}(g''(2y) + g''(0)). \tag{36}$$

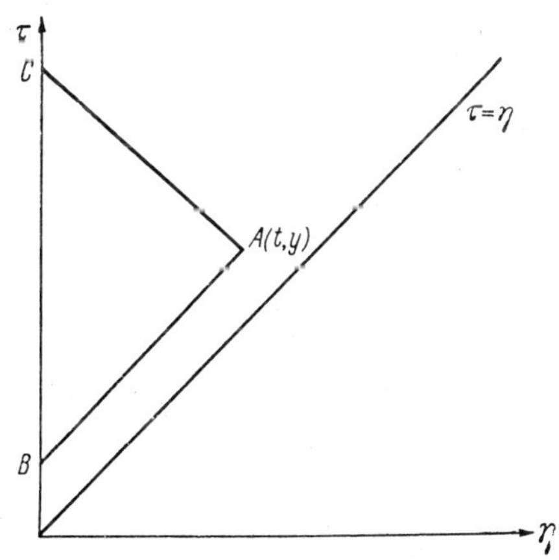

If instead of V and W we introduce the functions $V_0 = V$, $V_1 = V_t - V_y$, $V_2 = V_t + V_y$, $W_1 = W_t + W_y$ and $W_2 = W_t + W_y$ and the operators $L_t^{(1)}$, $L_t^{(2)}$ and M, where $L_t^{(1)}$ is the integration operator over the segment BA, $L_t^{(2)}$ is the operator over CA, and M is the operator over the triangle BAC (see the figure), then the equations for the functions V_i, W_i, A'_i and B'_2 may be written as follows:

$$V_0 = \frac{1}{4} M (XV_1 + ZV_2) + \frac{1}{2}[f(t+y) + f(t-y)]; \tag{37}$$

$$V_i = L_t^{(i)} (XV_1 + ZV_2) + f'[t + (-1)^i y], \quad i = 1,\ 2; \tag{38}$$

$$W_i = L_t^{(i)} (XW_1 + ZW_2 - 2C^2 V_0) + g'[t + (-1)^i y],\ i = 1,\ 2; \tag{39}$$

$$A'_0 = L_y^{(2)} (XV_1 + ZV_2) + f'(2y); \tag{40}$$

$$A'_1 = L_y^{(2)} (XV_{1t} + ZV_{2t}) + f''(2y); \tag{41}$$

$$B'_2 = \frac{1}{2} L_y^{(2)} \{XW_{1tt} + ZW_{2tt} + C^2 (V_1 + V_2)\} + \frac{1}{2} g'''(2y). \tag{42}$$

Thus we obtain a system of seven integro-differential equations for the functions $X(y)$, $Z(y)$, $V_i(y,t)$ and $W_i(y,t)$. The first five equations are equations (37)–(39). The last two are obtained by substituting (40)–(42) into formulas (27) and (28) for X and Z, allowing for (29) and (30). Here $L_y^{(2)} V$ is understood to mean $L_t^{(2)} V|_{t=y}$.

We may write this system in abbreviated form:

$$\begin{gathered} V_i(y,\ t) = K_i (V_1,\ V_2,\ X,\ Z),\ i = 0,\ 1,\ 2; \\ W_i(y,\ t) = K_{i+2} (V_0,\ V_1,\ V_2,\ W_1,\ W_2,\ X,\ Z),\ i = 1,\ 2; \\ X(y) = K_5 (V_1,\ V_2,\ X,\ Z); \\ Z(y) = K_6 (V_0,\ V_1,\ V_2,\ W_1,\ W_2,\ X,\ Z). \end{gathered}$$

§2. Investigation of the system of integro-differential equations. Existence and uniqueness theorem

In the preceding section we obtained a system of integro-differential equations for the functions V_0, V_1, V_2, W_1, W_2, X and Z; we write this system in the abbreviated form $\varphi = K\varphi$, where

$$\varphi = \begin{pmatrix} V_0 \\ V_1 \\ V_2 \\ W_1 \\ W_2 \\ X \\ Z \end{pmatrix},$$

and the operator K acts on the function φ as follows:

$$K\varphi = \begin{pmatrix} (K_0\varphi)(y, t) \\ (K_1\varphi)(y, t) \\ (K_2\varphi)(y, t) \\ (K_3\varphi)(y, t) \\ (K_4\varphi)(y, t) \\ (K_5\varphi)(y) \\ (K_6\varphi)(y) \end{pmatrix}$$

As regards K, we will show that in some Banach space L_{y0} there exists a ball C_r which is invariant with respect to K, and this operator is a contraction operator in this ball. Hence, according to the principle of contraction mappings, a unique solution belonging to L_{y0} exists in this ball. Thus we have proved the existence and uniqueness of the solution of our inverse problem over a finite depth z_0. The quantity z_0 is determined by the choice of L_{y0}. Consider a linear space L of the vector-valued functions φ:

$$\varphi = \begin{pmatrix} \varphi_0 (y, t) \\ \varphi_1 (y, t) \\ \varphi_2 (y, t) \\ \varphi_3 (y, t) \\ \varphi_4 (y, t) \\ \varphi_5 (y) \\ \varphi_6 (y) \end{pmatrix}$$

where $\varphi_0 \in C_{00}$, $\varphi_{1,2} \in C_{01}$, $\varphi_{3,4} \in C_{02}$, $\varphi_{5,6} \in C$. We introduce the norm

$$\|\varphi\| = \max\{\|\varphi_0\|_{C_{00}(\Delta)}, \|\varphi_1\|_{C_{01}(\Delta)}, \|\varphi_2\|_{C_{01}(\Delta)}, \|\varphi_3\|_{C_{02}(\Delta)}, \|\varphi_4\|_{C_{02}(\Delta)}, \|\varphi_5\|_{C[0, y_0]}, \|\varphi_6\|_{C[0, y_0]}\},$$

where Δ is the region $t \geqslant y$; $y \leqslant y_0$; $t + y \leqslant 2y_0$; y_0 is a quantity to be subsequently determined. Thus our linear space becomes a Banach space.

We will find a y_0 and r such that the following inequalities hold:

$$\|K\varphi\|_{L_{y_0}} \leqslant r, \tag{43a}$$

$$\|K\tilde{\varphi} - K\varphi\|_{L_{y_0}} \leqslant \alpha \|\tilde{\varphi} - \varphi\|_{L_{y_0}}, \quad \alpha < 1, \tag{43b}$$

if $\|\varphi\| \leqslant r$.

For this we estimate the norm of $K\varphi$. Without giving all the cumbersome calculations (though they are basically fairly simple), we will show how the estimate is obtained, e.g., for $\|K_1\varphi\|_{C_{01}(\Delta)}$:

$$\|K_1\varphi\|_{C_{01}(\Delta)} = \max\left\{\|K_1\varphi\|_{C_{00}(\Delta)}, \left\|\frac{\partial}{\partial t}K_1\varphi\right\|_{C_{00}(\Delta)}\right\},$$

$$\|K_1\varphi\|_{C_{00}(\Delta)} \leqslant \max|L_t^{(1)}(\varphi_5\varphi_1 + \varphi_6\varphi_2) + f'(t-y)|$$

$$= \max\left|\frac{1}{2}\int_0^y [\varphi_5(\eta)\varphi_1(\eta, t-(y-\eta)) + \varphi_6(\eta)\varphi_2(\eta, t-(y-\eta))]\,d\eta + f'(t-y)\right| \leqslant r^2y_0 + A. \tag{44}$$

$$A = \|f(t)\|_{C_2[0, 2y_0]}. \tag{45}$$

We similarly obtain the estimate

$$\left\|\frac{\partial}{\partial t}K_1\varphi\right\|_{C_{00}(\Delta)} \leqslant r^2y_0 + A;$$

thus

$$\|K_1\varphi\|_{C_{01}(\Delta)} \leqslant r^2y_0 + A.$$

By making estimates similar to (44), we arrive at the following assertion: if $\|\varphi\| < r$, then

$$\|K\varphi\| \leqslant \sum_{\substack{i,\,j,\,k \\ j+2\geqslant i}}^{k\leqslant 10;\; i,\,j\leqslant 4} \Phi_{ijk}\exp\left[\frac{kry_0}{2}\right]y_0^j r^i + G. \tag{46}$$

Here the Φ_{ijk} are polynomials with positive coefficients in A and B, where A is defined by (45), while $B = \|g(t)\|_{C_3[0,2y_0]}$, $G = \sum_{k=0}^{10}\exp[kry_0/2]G_k$. The G_k are polynomials in A and B.

Now consider (43b). We estimate, e.g., $\|K_1\tilde{\varphi} - K_1\varphi\|_{C_{01}(\Delta)}$:

$$\|K_1\tilde{\varphi} - K_1\varphi\|_{C_{00}(\Delta)} \leqslant \max\left|\frac{1}{2}\int_0^{y_0}[\tilde{\varphi}_5\tilde{\varphi}_1 + \tilde{\varphi}_6\tilde{\varphi}_2]\,d\eta - \frac{1}{2}\int_0^y[\varphi_5\varphi_1 + \varphi_6\varphi_2]\,d\eta\right|$$

$$\leqslant \frac{1}{2}\max\left[\int_0^{y_0}\{|\tilde{\varphi}_5||\tilde{\varphi}_1 - \varphi_1| + |\tilde{\varphi}_6||\tilde{\varphi}_2 - \varphi_2| + |\varphi_1||\tilde{\varphi}_5 - \varphi_5| + |\varphi_2||\tilde{\varphi}_6 - \varphi_6|\}\,d\eta\right] \leqslant 2ry_0\|\Delta\varphi\|;\quad \Delta\varphi = \tilde{\varphi} - \varphi.$$

A similar estimate yields

$$\|K_1\tilde{\varphi} - K_1\varphi\|_{C_{01}} \leqslant 2ry_0\|\Delta\varphi\|. \tag{47}$$

Making all the estimates, we arrive at the following assertion: if $\|\varphi\| \leqslant r$, then

$$\| K\tilde{\varphi} - K\varphi \| \leqslant \left[\sum_{\substack{i,j,k \\ j>i}} \Phi^{(2)}_{i,j,k} r^i y_0^i \exp\left(\frac{kry_0}{2}\right) + G^{(2)} \right] \| \Delta\varphi \|, \qquad (48)$$
$$G^{(2)} = \sum G^{(2)}_k \exp\left\{\frac{kry_0}{2}\right\},$$

where, as in (46), $\Phi^{(2)}_{ijk}$, and $G^{(2)}_k$ are polynomials with positive coefficients in A and B.

We will show that r and y_0 may be chosen such that (43a) and (43b) hold.

Since in our estimates $\|K_1 \tilde{\varphi} - K_1 \varphi\| \leqslant 2ry_0 \|\Delta\varphi\|$, we will assume that $ry_0 < 1/2$.

On the other hand, since $f(0) = -C(0)$ and hence $V(0,0) = -C(0)$, it follows that $\|V\|$ is bounded from below in terms of $|C(0)|$. Therefore we must take $r > |C(0)|$, which implies that $y_0 < 1/2|C(0)|$. Hency by taking y_0 to be such that $G^{(2)} < 1/2$, i.e. by satisfying the inequality $I = \sum_k G^{(2)}_k \exp\{k/4\} y_0 < 1/2$, we can find an n such that inequality (43b) holds for $ry_0 < 1/n$. This may be done because all the terms in (48) are estimated in terms of vy_0.

All the terms in (46) are estimated in terms of ry_0 and $r^2 y_0$. Using the estimate $ry_0 < 1/n$, we find that

$$G < \sum G_k \exp\left\{\frac{k}{2n}\right\} = H.$$

We take $r = 2H$ and choose a y_0 such that the following inequalities hold:

$$y_0 < \frac{1}{nr} \mp \frac{1}{2Hn}, \qquad y_0 < \frac{1}{2I}, \qquad y_0 < \frac{1}{4H^2 m}, \qquad (49)$$

where m is introduced for (46) in a manner similar to n.

It is clear that this may always be done. Then inequalities (43a) and (43b) will be simultaneously satisfied, which means that in the ball $C_r = C_{2H}$ of the Banach space L_{y0} our solution exists and is unique.

Knowing $X(y)$ and $Z(y)$ from (19) and (32), we can find $\sigma(y)$ and $C(y)$ for $y \leqslant y_0$.

Now, using (9), we can find $C(z)$ and $\sigma(z)$. Thus we have solved our problem.

Bibliography

1. M.G. Kreĭn, *On the transfer function of a one-dimensional boundary problem of second order*, Dokl. Akad. Nauk SSSR **88** (1953), 405–408. (Russian) MR **15**, 316.

2. A.S. Blagoveščenskiĭ, *Am inverse problem in the theory of propagation of seismic waves*, Problemy Mat. Fiz., no. 1, Izdat. Leningrad. Univ., Leningrad, 1966, pp. 68–81; English transl., Topics in Math. Phys., no. 1, Consultants Bureau, New York, 1967, pp. 55–67.

Translated by:
M. Lopez-Morillas

THE FIELD OF A POINT SOURCE IN A WAVEGUIDE

V. S. BULDYREV

UDC 517 944/947

Abstract. Suppose that in the xy-plane we are given a function $c(x,y)$ that has the physical meaning of the velocity of wave propagation. We shall consider the problem of the wave field of a point source located near the line of minimum of the velocity c. First we study the case

$$c(y) = c_0 [1 - 2ny^2]^{-1/2},$$

for which the problem admits an explicit solution. From the latter the author was able to separate a so-called axial wave that propagates along the "waveguide axis," i.e. near the straight line $y = 0$. In the last section these results are extended to a more general case of velocities $c(x,y)$.

Figures: 8. Bibliography: 5 items

This paper is devoted to a study of the short-wave asymptotic behavior of the two-dimensional Green's function for Helmholtz's equation in the case of an inhomogeneous medium, on the assumption that there exists in the medium a layer of reduced velocity of wave propagation. If the source is located on the line of minimum of the velocity $c(x,y)$ (on the waveguide axis) in the xy-plane, then the rays[1] emerging from the source at sufficiently small angles will intersect infinitely many times the line of minimum. In the vicinity of the line of minimum there exist a large number of caustics (ray envelopes), each of which has a turning point on the line of minimum and consists of two branches. The interference of the wave fields corresponding to the rays, located in the vicinity of the caustics near the line of minimum, produces a special wave that propagates along this line. In the following this special wave will be called an "axial wave."

Axial waves are of considerable interest in various applied problems (marine acoustics, seismology, and the theory of radio wave propagation in the ionosphere). It is precisely this type of wave that we investigate in this paper.

In the first and the second section we separate the axial wave and study it in a "reference problem" in which there is a special relationship between the velocity $c(x,y)$ and the coordinates, such that the solution of Helmholtz's equation can be expressed in terms of functions of a parabolic cylinder. For the axial wave we obtain a formula containing a contour integral which is a special function $V_N(\xi,\eta)$ that must be tabulated.

In the third section we obtain a formula for the axial wave in the case of a

AMS 1970 *subject classifications.* Primary 35J05, 78A40; Secondary 76Q05, 86A15.

[1] Here and in the sequel we shall understand by rays the extremals of the integral $\int (1/c(x,y))ds$.

symmetrical waveguide of arbitrary cross-section. This formula is a linear superposition consisting of solutions of the wave equation that are concentrated in the vicinity of the line of minimum. Such solutions were constructed by the author in an earlier paper [1]. It is found that if the axial wave leads the waves corresponding to rays emerging from the source at sufficiently large angles, it will be described by a function $V_N(\xi,\eta)$. If, on the other hand, the axial wave lags behind such waves, it will be described by a function that differs only insignificantly from $V_N(\xi,\eta)$.

If the frequency ω or the coordinate of the observation point increases, it is possible to obtain simpler expressions from the special function describing an axial wave; these expressions represent geometrical waves propagating in the system under consideration. The intensity of such waves can be calculated also with the aid of ordinary formulas of the ray method.

§1. Reference problem. Field of concentrated source in a medium with a special velocity of wave propagation

We shall assume that the velocity of wave propagation in the medium depends only on one coordinate, being described by the function

$$c(y)=c_0[1-2ny^2]^{-1/2}, \tag{1.1}$$

where c_0 and n are positive constants. Together with the velocity of wave propagation, we shall use also the relative refractive index

$$n(y)=\frac{c_0}{c(y)}=\sqrt{1-2ny^2} \tag{1.2}$$

which is the reciprocal of (1.1).

The functions $c(y)$ and $n(y)$ are real in the interval $[-1/\sqrt{2n}, 1/\sqrt{2n}]$ and have a unique extremum ($c(y)$ has a minimum, and $n(y)$ a maximum) at the point $y=0$. At a velocity of wave propagation described by the function (1.1), the wave motion can take place only in the layer $-1/\sqrt{2n}<y<1/\sqrt{2n}$. Outside this layer the wave equation will be of elliptic type, and there can be no propagation of wave fronts.

In the following we shall consider the two-dimensional problem, when the wave field depends only on two spatial coordinates x and y (a linear source). The time multiplier $e^{-i\omega t}$, where ω is the oscillation frequency, will not be written out.

1. The reference problem for the field of a concentrated source is formulated as follows. We shall denote $\omega/c_0=k$, and assume k to be a complex quantity $k=k'+ik''$ with a small positive imaginary part $k''>0$. Let $x=0$ and $y=y_0$ ($|y_0|<1/\sqrt{2n}$) be the coordinates of the source. The wave field for $y<y_0$ will be denoted by $u_1(x,y)$, whereas for $y>y_0$ it will be denoted by $u_2(x,y)$. The functions $u_j(x,y)$, $j=1, 2$, must satisfy Helmholtz's equation

$$\frac{\partial^2 u_j}{\partial x^2}+\frac{\partial^2 u_j}{\partial y^2}+k^2(1-2ny^2)u_j=0, \tag{1.3}$$

the conditions of matching "at the horizon" of the source

$$u_2(x, y_0) = u_1(x, y_0),$$
$$\left.\frac{\partial u_2}{\partial y}\right|_{y=y_0} - \left.\frac{\partial u_1}{\partial y}\right|_{y=y_0} = -2\pi\delta(x) \tag{1.4}$$

and the condition of absence of a wave field at infinity

$$u_j(x, y) \to 0 \quad \text{for} \quad r = \sqrt{x^2 + y^2} \to \infty. \tag{1.5}$$

In (1.4), $\delta(x)$ is Dirac's function. The solution of the problem (1.3)–(1.5) can be constructed by the method of separation of variables, and it has the form [2]

$$u_j(x, y) = k \int_{-\infty}^{\infty} \frac{\Phi_l(y_0, \zeta)\,\Phi_j(y, \zeta)}{W(\zeta)} e^{ikx\zeta} d\zeta, \qquad j, l = 1, 2;\ l \neq j, \tag{1.6}$$

where the $\Phi_j(y, \zeta)$ are linearly independent solutions of the equation

$$\Phi_1(y, \zeta) \to 0 \quad \text{for} \quad y \to -\infty, \qquad \Phi_2(y, \zeta) \to 0 \quad \text{for} \quad y \to \infty,$$

satisfying the conditions

$$\frac{d^2\Phi}{dy^2} + k^2[1 - 2ny^2 - \zeta^2]\,\Phi = 0, \tag{1.7}$$

and $W(\zeta)$ is the Wronskian

$$W(\zeta) = \Phi_1(y, \zeta)\,\Phi_2'(y, \zeta) - \Phi_1'(y, \zeta)\,\Phi_2(y, \zeta).$$

For the convergence of the integral in (1.6), the integration contour must be raised for $x > 0$ to the second quadrant of the ζ-plane, and lowered to the fourth quadrant for $x < 0$. In the following we shall consider only $x \geqslant 0$, since the wave field $u_j(x,y)$ is an even function of x. The functions $\Phi_1(y, \zeta)$ and $\Phi_2(y, \zeta)$ can be expressed in terms of parabolic-cylinder functions $D_t(\pm\sqrt{2}\,\nu)$ that have the following behavior for $|\nu| \to \infty$:

$$D_t(\sqrt{2}\,\nu) \sim e^{-\frac{\nu^2}{2}} \nu^t \left\{2^{t/2} - \frac{t(t-1)}{4\nu^2} + \ldots\right\}, \quad |\arg\nu| < \frac{3\pi}{4},$$

$$D_t(-\sqrt{2}\,\nu) \sim e^{-\frac{\nu^2}{2}} (-\nu)^t \left\{2^{t/2} - \frac{t(t-1)}{4\nu^2} + \ldots\right\}, \quad |\pi - \arg\nu| < \frac{3\pi}{4}.$$

The formulas connecting $\Phi_j(y, \zeta)$ and $D_t(\pm\sqrt{2}\,\nu)$ are

$$\Phi_2(y, \zeta) = D_{\frac{k}{2\sqrt{2n}}(1-\zeta^2) - \frac{1}{2}}\left(\sqrt[4]{8nk^2}\,y\right), \tag{1.8}$$

$$\Phi_1(y, \zeta) = D_{\frac{k}{2\sqrt{2n}}(1-\zeta^2) - \frac{1}{2}}\left(-\sqrt[4]{8nk^2}\,y\right).$$

2. Let us examine the distribution of the singular points of the integrand in the ζ-plane. We shall calculate the Wronskian $W(\zeta)$ by expanding the function $D_t(\sqrt{2}\,\nu)$ in a Maclaurin series:

$$D_t(\sqrt{2}\nu) = \sqrt{\pi}2^{t/2}\left\{\frac{1}{\Gamma\left(\frac{1-t}{2}\right)} - \frac{2}{\Gamma\left(-\frac{t}{2}\right)}\nu + O(\nu^2)\right\}, \qquad (1.9)$$

where $\Gamma(z)$ is Euler's gamma function. Using the doubling formulas for the gamma function, we obtain

$$D_t(\sqrt{2}\nu)\,D_t'(-\sqrt{2}\nu) - D_t'(\sqrt{2}\nu)\,D_t(-\sqrt{2}\nu) = \frac{2}{\sqrt{\pi}}\Gamma(t+1)\sin\pi t$$

and hence

$$W(\zeta) = \frac{2\sqrt[4]{2nk^2}}{\sqrt{\pi}}\Gamma(t+1)\sin\pi t, \quad t = \frac{1}{2}\left[\frac{k}{\sqrt{2n}}(1-\zeta^2) - 1\right]. \qquad (1.10)$$

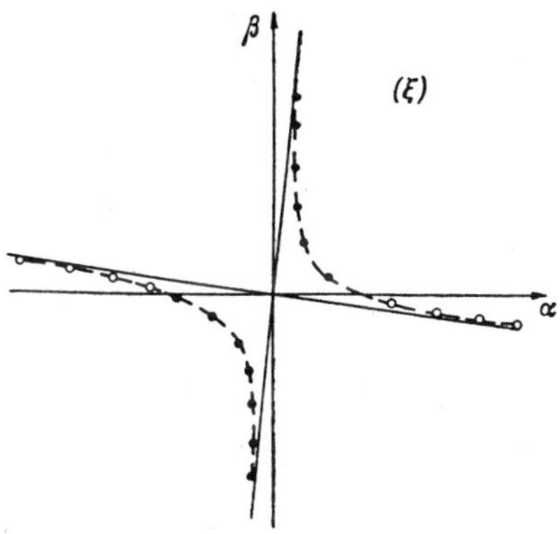

Figure 1.

Since parabolic-cylinder functions are entire functions in the t-plane, the integrand in (1.6) will have simple poles only at the roots of the equation $\Gamma(t+1)\sin\pi t = 0$ or $\Gamma^{-1}(-t) = 0$, i.e. for $t = m = 0, 1, \ldots$. The poles in the ζ-plane can be determined from the equation

$$\frac{1}{2}\left[\frac{k}{\sqrt{2n}}(1-\zeta^2) - 1\right] = m,$$

$$m = 0,\ 1,\ 2,\ \ldots,$$

whence $\zeta = \pm\zeta_m \equiv \pm(1 - (2m+1)\sqrt{2n}/k)^{1/2}$. Thus the poles $\pm\zeta_m$ will lie in the ζ-plane symmetrical with respect to the origin of coordinates. With real k and $m < \frac{1}{2}(k/\sqrt{2n} - 1) \equiv m_0$, the poles $\pm\zeta_m$ will be real, whereas for $m > m_0$ they will be purely imaginary. For complex values of $k = k' + ik''$ the poles $\pm\zeta_m$ will move into the complex plane $\zeta = \alpha + i\beta$, where they are located in the first and third quadrants on the lines

$$\operatorname{Im} k(1-\zeta^2)=0. \tag{1.11}$$

In Figure 1 the lines (1.11) are plotted in the form of dashed lines representing rectangular hyperbolas with asymptotes

$$\beta=\cot\left(\frac{1}{2}\arg k\right)\alpha,$$

$$\beta=-\tan\left(\frac{1}{2}\arg k\right)\alpha.$$

In the figure, the poles are represented by dots.

By using asymptotic formulas for parabolic-cylinder functions that describe the behavior of the latter in the complex t-plane, it is possible to show that the ratio

$$\zeta\frac{\Phi(y_0,\zeta)\,\Phi(y,\zeta)}{W(\zeta)},$$

remains (apart from the poles $\pm\zeta_m$) bounded in the ζ-plane, including the point at infinity. Hence for $x>0$ we can represent $u_j(x,y)$ by a sum of the residues of the integrand function at the poles $+\zeta_m$ lying in the upper halfplane, whereas for $x<0$ it can be represented by a sum of the residues at the poles $-\zeta_m$ lying in the lower halfplane. For $x\to\infty$ the residues at the poles ζ_m tend to zero, and for $x\to-\infty$ the residues at the poles $-\zeta_m$ tend to zero. Thus the wave fields $u_j(x,y)$, $j=1,2$, satisfy the condition $\lim_{|x|\to\infty}u_j(x,y)=0$ (from (1.6) it follows that $u_j(x,y)$ tends to zero for $|y|\to\infty$). The residues at the poles $\pm\zeta_m$, $m<m_0$, yield so-called propagating waveguide waves, whereas the residues at the poles $\pm\zeta_m$, $m>m_0$, describe decaying waves.

To represent the wave field by a sum of propagating and decaying guide waves amounts in fact to expanding the solution into a discrete spectrum. Such a representation does not correspond to the concept of rays and is hardly compatible with the formulas of the ray method.

Let us now transform the exact solution (1.6) into a sum whose terms could be interpreted for $\omega\to\infty$ on the basis of the ray method. First we shall consider the geometrical pattern of rays occurring in such a problem.

3. Since the velocity of wave propagation (1.1) does not depend on the x-coordinate, Euler's equation for rays will admit in this problem a first integral

$$\sqrt{1-2ny^2}=c_1\sqrt{1+y'^2}. \tag{1.12}$$

The solution of equation (1.12), satisfying the conditions

$$y(0)=y_0\quad\left(|y_0|<\frac{1}{\sqrt{2n}}\right),\quad y'(0)=\tan\psi,$$

where ψ is the angle of exit of the ray from the source, i.e. the angle of inclination of the tangent to the ray at the point $x=0$, $y=y_0$, will be

$$y = \sin\psi \sqrt{\frac{1}{2n} - y_0^2}\, \sin \frac{x'}{\cos\psi \sqrt{\frac{1}{2n} - y_0^2}} + y_0 \cos \frac{x}{\cos\psi \sqrt{\frac{1}{2n} - y_0^2}}. \quad (1.13)$$

With given coordinates of the observation point x and y ($|y| < 1/\sqrt{2n}$), the equation (1.13) will have infinitely many solutions in ψ. This signifies that infinitely many rays will arrive at the observation point that have left the source at different angles ψ. Let us write $y_0 = y = 0$; hence the solutions of (1.13) will be

$$\psi = 0 \quad \text{and} \quad \cos\psi = \zeta_j = \frac{l}{\pi j}, \quad (1.14)$$

where $l = x/(1/\sqrt{2n})$ is the dimensionless coordinate of the observation point, and j is an integer that must satisfy the inequality

$$j > \frac{l}{\pi}, \quad (1.15)$$

since $0 \leqslant \zeta_j = \cos\psi \leqslant 1$. The quantity $a = 1/\sqrt{2n}$, which has the dimension of length, is very important in an investigation of the wave field near the waveguide axis; the quantity a will be called by us the characteristic length of the waveguide.

By substituting (1.4) into (1.13), we obtain the entire set of rays arriving at the observation point. In dimensionless coordinates

$$\tilde{y} = \frac{1}{a} y \text{ and } \tilde{x} = \frac{x}{a} \quad (1.16)$$

this set will be written as

$$\tilde{y} = 0, \quad \tilde{y} = \sqrt{1 - l^2 (\pi j)^{-2}} \sin\left(\pi j \frac{\tilde{x}}{l}\right), \quad j > \frac{l}{\pi}.$$

The number j evidently shows how many times the ray L crosses the waveguide axis on its way from the source to the observation point (the source together with the observation point are regarded as a single crossing point).

Let L_j be a ray that arrives at the observation point after j crossings of the waveguide axis. Inequality (1.15) signifies that an observation point situated in such a way that $\pi N \leqslant l < \pi(N + 1)$, $N = 0, 1, 2, \ldots$, will receive rays L_j such that $j \geqslant N + 1$, i.e. rays that cross the waveguide axis not less than $N + 1$ times.[2] Let us denote by U_j, $j \geqslant 1$, a wave corresponding to the ray L_j, and by U_0 a wave that propagates along the waveguide axis. Let us calculate the propagation time T_j of a perturbation along a ray L_j that leaves the source at an angle $\psi_j = \arccos \zeta_j$ and arrives at the observation point after j crossings of the waveguide axis. Evidently

[2]) This represents an essential difference between a "depth waveguide" ($c' = 0$) and a "surface waveguide" ($c' > 0$).

$$T_j = \frac{1}{c_0}\int_0^x \sqrt{1-2ny^2(x)}\,\sqrt{1+y'^2(x)}\,dx = \frac{\pi}{2}\tau_0\left[1+\left(\frac{l}{\pi j}\right)^2\right]j, \quad (1.17)$$

where $\tau_0 = a/c_0$ is the characteristic time of the waveguide.

It is natural to assume that two waves with numbers j_2 and j_1 can be observed at a point $(\ell, 0)$ independently from one another if their arrival times satisfy the inequality

$$T_{j_2} - T_{j_1} > q\,(kal)^{2\delta}\,T, \quad (1.18)$$

where

$$T = \frac{2\pi}{\omega} = 2\pi\frac{\tau_0}{ak}, \quad q \sim 1 \div 3, \; 0 < \delta < 1.$$

With the aid of the factor $(ka\ell)^{2\delta}$ in the right-hand side of (1.18) we shall obtain below estimates that tend to zero when $ka \to \infty$. Let us note that for moderate values of the large parameter ka and small δ, the value of $(ka\ell)^{2\delta}$ will be close to unity. Let us find the difference $T_j - T_0$ between the times of arrival at the observation point $(\ell, 0)$ of a wave with number j and an axial wave U_0. Using (1.17), we obtain

$$T_j - T_0 = \frac{\pi}{2}\tau_0\left[1+\left(\frac{l}{\pi j}\right)^2\right]j - \tau_0 l = \frac{\pi}{2}\tau_0\left(j-\frac{l}{\pi}\right)^2\frac{1}{j}. \quad (1.19)$$

By substituting (1.19) into (1.18) and neglecting terms of the order of $(ka)^{-1+2\delta}$, we obtain the condition

$$j > \frac{l}{\pi}\left[1+\Delta\,(kal)^{-1/2+\delta}\right], \quad \Delta = 2\sqrt{\pi q} = O(1), \quad (1.20)$$

under which the waves U_j and U_0 can be observed independently. But if

$$\frac{l}{\pi} < j \leqslant \frac{l}{\pi}\left[1+\Delta\,(kal)^{-1/2+\delta}\right], \quad (1.21)$$

these waves will have interference, causing the formation of a wave field of fairly complex structure. It follows from condition (1.21) that with small ℓ such interference will take place only between waves U_N and U_0 in a neighborhood of points $\ell = \pi N$, $N = 1, 2, \ldots$, of the order of $(ka)^{-1/2+\delta}$. This neighborhood increases with ℓ.

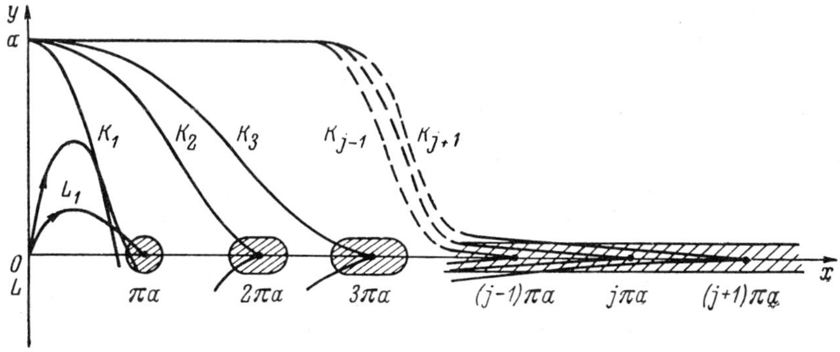

Figure 2.

In the case of large $\ell \sim \pi ka$, we can have interference of several waves U_j with the wave U_0. The region in which this interference takes place will be of the order of several characteristic lengths. Thus whereas at short distances from the source the formulas of the ray method are applicable except in individual "gaps" of the waveguide axis, these gaps will merge in the case of large distances $\ell \sim \pi ka$, and the axial wave U_0 assumes a complex interference character. With increasing distance ℓ the axial wave U_0 will absorb waves of ever greater number. It follows from condition (1.21) that a wave U_j loses its individuality during the interference process beginning with distances

$$l \sim \pi j [1 - \Delta (ka\pi j)^{-1/2+\delta}], \quad j \sim ka.$$

Let us note that the instant of absorption of a wave U_j by an axial wave is arbitrary, i.e. it depends on the choice of the coefficient q in (1.18). With $\ell \gg \pi ka$ we also have mutual interference of waves U_j. Using the formula (1.17), it is easy to see that with a fixed observation point $(\ell, 0)$ we shall have mutual interference of waves U_j such that

$$\frac{l}{\pi} < j \leqslant \frac{l}{\pi}[1 + 2\pi q l (kal)^{-1+2\delta}]. \tag{1.22}$$

In Figure 2 we plotted the caustics K_j for rays that cross the waveguide axis j times, and the neighborhoods of points πN at which interference of various waves takes place are hatched.

§2. Separation of axial wave from exact solution

Having established certain geometrical laws that hold in our problem, let us return to the study of the exact solution (1.6). The aim of the subsequent transformations of the solution (1.6) is to separate formulas from this solution which for $\omega \to \infty$ describe mutually noninterfering waves U_j, as well as an axial wave U_0 that propagates along the waveguide axis.

1. In the numerator of (1.6) we go over from real solutions $D_t(\sqrt{2}\,\nu)$ and $D_t(-\sqrt{2}\,\nu)$ of equation (1.7) to complex-valued solutions $d_t^{(1)}(\nu)$ and $d_t^{(2)}(\nu)$. The functions $d_t^{(1)}(\nu)$ and $d_t^{(2)}(\nu)$ are defined by the formulas

$$d_t^{(1)}(\nu)=\sqrt{\frac{2}{\pi}}\,\Gamma(t+1)\,e^{i\frac{\pi}{2}(t+1)}D_{-t-1}(i\sqrt{2}\nu), \tag{2.1}$$

$$d_t^{(2)}(\nu)=\sqrt{\frac{2}{\pi}}\,\Gamma(t+1)\,e^{-i\frac{\pi}{2}(t+1)}D_{-t-1}(-i\sqrt{2}\nu).$$

It follows from (2.1) that

$$d_t^{(1)}(-\nu)=-e^{i\pi t}d_t^{(2)}(\nu) \tag{2.2}$$

and

$$d_t^{(2)}(-\nu)=-e^{-i\pi t}d_t^{(1)}(\nu).$$

By virtue of the equation

$$D_t(\sqrt{2}\nu)=\frac{\Gamma(t+1)}{\sqrt{2\pi}}\left\{e^{i\frac{\pi}{2}t}D_{-t-1}(i\sqrt{2}\nu)+e^{-i\frac{\pi}{2}t}D_{-t-1}(-i\sqrt{2}\nu)\right\}$$

we obtain

$$D_t(\sqrt{2}\nu)=\frac{1}{2i}(d_t^{(1)}(\nu)-d_t^{(2)}(\nu)). \tag{2.3}$$

Using the formulas (1.10) and (2.3), for $y \geqslant y_0$ we can write the exact solution (1.16) in dimensionless coordinates as follows:

$$u_2(l,\ \tilde{\nu};\ \tilde{\nu}_0)$$

$$=-\frac{\sqrt{\pi}}{8}\sqrt{ka}\int_{-\infty}^{\infty}\frac{[d_t^{(1)}(-\tilde{\nu}_0)-d_t^{(2)}(-\tilde{\nu}_0)][d_t^{(1)}(\tilde{\nu})-d_t^{(2)}(\tilde{\nu})]}{\Gamma(t+1)\sin\pi t}\,e^{ikal\zeta}d\zeta, \tag{2.4}$$

where $\tilde{\nu}_0=\sqrt{ka}\,y_0/a$ is the dimensionless "reduced height" of the source, and $\nu=\sqrt{ka}\,y/a$ the "reduced height" of the observation point. For $y \leqslant y_0$ the wave field $u_1(l,\tilde{\nu};\nu_0)$ can be obtained from (2.4) by permutation of $\tilde{\nu}_0$ and $\tilde{\nu}_1$.

Below we shall study in detail the solution (2.4) in the particular case that the wave-field source and the observation point lie on the waveguide axis, i.e. $\tilde{\nu}_0=\tilde{\nu}=0$. In the general case $\tilde{\nu}_0\neq\tilde{\nu}\neq 0$, it is possible to study the solution (2.4) by similar methods. By setting $\tilde{\nu}_0=\tilde{\nu}=0$ in (2.4), and using the formulas (2.1) and (2.9), we obtain

$$u(l,\ 0;\ 0)=i\frac{\sqrt{\pi}}{4}\sqrt{ka}\int_{-\infty}^{\infty}2^{-t}\frac{\Gamma(t+1)}{\Gamma^2\left(\frac{t}{2}+1\right)}\,\frac{1+e^{i\pi t}}{1-e^{i\pi t}}\,e^{ikal\zeta}\,d\zeta. \tag{2.5}$$

With the aid of Stirling's formula

$$\ln\Gamma(z)=z\ln z-z-\frac{1}{2}\ln z+\ln\sqrt{2\pi}+0\left(\frac{1}{z}\right),\quad |\arg z|<\pi-\varepsilon,\quad \varepsilon>0,$$

we can establish the behavior of the factor

$$2^{-t}\frac{\Gamma(t+1)}{\Gamma^2\left(\frac{t}{2}+1\right)}$$

in the t-plane:

$$2^{-t}\frac{\Gamma(t+1)}{\Gamma^2\left(\frac{t}{2}+1\right)}=\begin{cases}\sqrt{\frac{2}{\pi t}}\left[1+0\left(\frac{1}{t}\right)\right], & |\arg t|<\pi-\varepsilon.\\ i\sqrt{\frac{2}{\pi t}}\tan\frac{\pi t}{2}\left[1+0\left(\frac{1}{t}\right)\right], & \varepsilon<\arg t<2\pi-\varepsilon.\end{cases} \tag{2.6}$$

By going over in the first relation (2.6) to the variable ζ, we obtain

$$2^{-t}\frac{\Gamma(t+1)}{\Gamma^2\left(\frac{t}{2}+1\right)}=\frac{2}{\sqrt{ka(1-\zeta^2)}}\left[1+0\left(\frac{1}{ka(1-\zeta^2)}\right)\right]. \tag{2.7}$$

Formula (2.7) holds in the entire ζ-plane, apart from sectors enclosing parts of the lines (1.11) located in the second and the fourth quadrant (for real k the sectors enclose the rays $\operatorname{Re}\zeta < -1$ and $\operatorname{Re}\zeta > 1$). In the sectors themselves the formula for the factor (2.5a) can be obtained from the second relation (2.6); it contains $\tan[ka/4)(1-\zeta^2)-\pi/4]$. Hence the factor (2.5a) has simple poles on the lines (1.11) in the second and the fourth quadrant. These poles are marked by small circles in Figure 1. The form of the remainder in (2.7) shows that this formula cannot be used in a neighborhood of the points $\zeta=\pm 1$.

2. For further transforming the solution (2.5), it is necessary to study the behavior of the function

$$F(\zeta)=e^{i\pi Nt+ikal\zeta} \tag{2.8}$$

in the complex plane $\zeta=\alpha+i\beta$. By substituting into (2.8) the value of the function $t(\zeta)$, we obtain

$$F(\zeta)=e^{-i\frac{\pi}{2}N}\cdot e^{ipf_N(\zeta)},$$

where

$$p=\pi|ka|\frac{N}{2},\quad f_N(\zeta)=\left[1+\frac{l^2}{\pi^2N^2}-\left(\zeta-\frac{l}{\pi N}\right)^2\right]e^{i\varkappa},\quad \varkappa=\arg k.$$

The phase function $f_N(\zeta)$ has on the ζ-plane a unique saddle point $\zeta_N = l/\pi N$ $(f'_N(\zeta_N)=0)$.

The lines of constant modulus $F(\zeta)$ passing through the saddle point are described by the equation

$$\operatorname{Im} f_N(\zeta) = \operatorname{Im} f_N(\zeta_N) = \left(1 + \frac{l^2}{\pi^2 N^2}\right) \sin \varkappa$$

they are in the form of two straight lines:

$$\beta = \cot \frac{\varkappa}{2}\left(\alpha - \frac{l}{\pi N}\right), \qquad \beta = -\tan \frac{\varkappa}{2}\left(\alpha - \frac{l}{\pi N}\right). \tag{2.9}$$

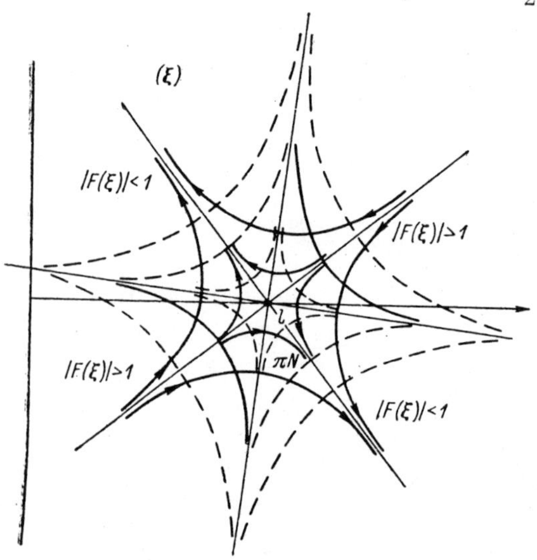

Figure 3

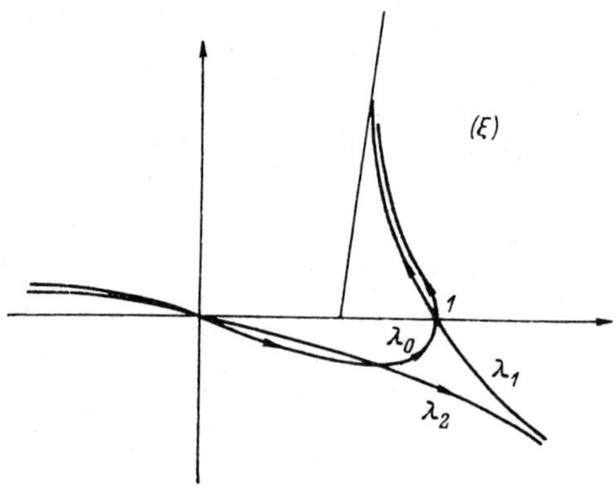

Figure 4

The other lines $|F(\zeta)| = \text{const}$ form a family of rectangular hyperbolas for which the straight lines (2.9) are asymptotes. In Figure 3 these lines are dashed. The thick solid lines represent the branches of the hyperbola $\operatorname{Im} f_N(\zeta) = 0$ on which $|F(\zeta)| = 1$. These lines separate the regions of the ζ-plane with $|F(\zeta)| > 1$ and $|F(\zeta)| < 1$. The family of orthogonal curves $\operatorname{Re} f_N(\zeta) = \text{const}$ of fastest variation of $|F(\zeta)|$ are plotted in Figure 3 by thin solid lines. The arrows indicate the direction of decreasing $|F(\zeta)|$.

The first step in the transformation of the solution (2.5) involves the deformation of the integration contour. In the ζ-plane we construct a straight line

$$\beta = \cot \frac{\varkappa}{2}(\alpha - \tilde{\alpha}), \tag{2.10}$$

where $0 < \tilde{a} < l/\pi N$, N being an integer, with $\pi N > l$. The exact value of N will be indicated later. Without crossing the poles of the integrand, we shall deform the integration contour in (2.5) in such a way that the new contour λ_0 will lie in the second quadrant higher than the pole line (1.11), then pass through the point $\zeta = 1$, and go to infinity in the first quadrant along the straight line (2.10) (Figure 4).

Such a deformation of the integration contour is possible owing to the decrease of the function $\exp(ika\ell\zeta)$, $\ell > 0$, in the upper halfplane. On the integration contour λ_0 we shall expand the function $(1 + \exp(i\pi t))/(1 - \exp(i\pi t))$ into a geometrical progression:

$$\frac{1+e^{i\pi t}}{1-e^{i\pi t}}=1+2\sum_{j=1}^{N-1} e^{i\pi tj}+2\frac{e^{i\pi tN}}{1-e^{i\pi t}}. \tag{2.11}$$

The integrals

$$\int_{\lambda_0} 2^{-t}\frac{\Gamma(t+1)}{\Gamma^2\left(\frac{t}{2}+1\right)} e^{i\pi tj+ikal\zeta}\,d\zeta$$

vanish for $j = 0, 1, \ldots, N - 1$, since the integrand does not have singularities in the region to the left of the integration contour λ_0 and it tends sufficiently fast to zero when $|\zeta| \to \infty$ in this region. In the upper halfplane the last term of (2.11) does not exceed $|\exp[i\pi t(N - 1)]|$, and therefore the integral of this term along the contour λ_0 will be convergent.

Hence

$$u(l,\ 0;\ 0)=i\frac{\sqrt{\pi}}{2}(ka)^{1/2}\int_{\lambda_0} 2^{-t}\frac{\Gamma(t+1)}{\Gamma^2\left(\frac{t}{2}+1\right)}\frac{e^{i\pi tN}}{1-e^{i\pi t}}e^{ikal\zeta}\,d\zeta. \tag{2.12}$$

In the sector

$$\frac{3\pi}{2}<\arg\zeta<2\pi-\frac{\varkappa}{2} \tag{2.13}$$

the integrand of (2.12) will behave as $\exp[i\pi N + ika\ell]$.

For $N \geqslant 1$ the function $\exp[i\pi tN + ika\ell]$ will tend to zero when $|\zeta| \to \infty$ within the sector (2.13). In (2.12) let us replace one integral along the contour λ_0 by two integrals taken over the contours λ_1 and λ_2. The contour λ_1 begins at infinity in the sector (2.13), and in the first quadrant it coincides with the contour λ_0; the contour λ_2 coincides with the contour λ_0 in the second quadrant, and it goes to infinity in the sector (2.13) (Figure 4). Hence $u(\ell, 0; 0) = I_N(\ell) + H_N(\ell)$, where

$$I_N(l)=i\frac{\sqrt{\pi}}{2}(ka)^{1/2}\int_{\lambda_2} 2^{-t}\frac{\Gamma(t+1)}{\Gamma^2\left(\frac{t}{2}+1\right)}\frac{e^{i\pi tN+ikal\zeta}}{1-e^{i\pi t}}\,d\zeta, \tag{2.14}$$

$$H_N(l)=i\frac{\sqrt{\pi}}{2}(ka)^{1/2}\int_{\lambda_1} 2^{-t}\frac{\Gamma(t+1)}{\Gamma^2\left(\frac{t}{2}+1\right)}\frac{e^{i\pi tN+ikal\zeta}}{1-e^{i\pi t}}\,d\zeta.$$

3. In I_N we transform $e^{i\pi tN}(1 - e^{i\pi t})^{-1}$ as follows:

$$\frac{e^{i\pi tN}}{1-e^{i\pi t}}=\sum_{m=0}^{M-1} e^{i\pi t(N+m)}+\frac{e^{i\pi t(N+M)}}{1-e^{i\pi t}}. \tag{2.15}$$

With the aid of the expansion (2.15) it is possible to express the integral I_N by a sum of integrals:

$$I_N=\sum_{m=0}^{M-1} s_{N+m}+S_{N+M}, \tag{2.16}$$

where

$$s_{N+m}=i\frac{\sqrt{\pi}}{2}(ka)^{1/2}\int_{\lambda_2} 2^{-t}\frac{\Gamma(t+1)}{\Gamma^2\left(\frac{t}{2}+1\right)}e^{i\pi t(N+m)+ikal\zeta}\,d\zeta \tag{2.17}$$

and

$$S_{N+M}=i\frac{\sqrt{\pi}}{2}(ka)^{1/2}\int_{\lambda_2} 2^{-t}\frac{\Gamma(t+1)}{\Gamma^2\left(\frac{t}{2}+1\right)}\frac{e^{i\pi t(N+M)+ikal\zeta}}{1-e^{i\pi t}}\,d\zeta. \tag{2.18}$$

The integrals s_{N+m} can be evaluated by the method of steepest descent at the saddle points $\zeta_{N+m} = l[\pi(N + m)]^{-1}$ of the phase function $f_{N+m}(\zeta) = i\pi t(N + m) + ikal$. As we have seen, in the ζ-plane (apart from sectors enclosing the lines of poles in the second and fourth quadrant) the function (2.5a) is a slowly varying function that satisfies the asymptotic formula (2.7). With increasing l the saddle points ζ_{N+m} move along the real axis from left to right, whereas for increasing m they move from right to left. Under the condition $N > l/\pi$ all the saddle points ζ_{N+m}, $m = 0, 1, \ldots,$ will lie on the segment $0 < \zeta < 1$, where the asymptotic formula (2.7) can be used. With the aid of (2.7) and of ordinary calculations used in the method of steepest descent, we obtain

$$\begin{aligned} s_{N+m}&=\sqrt{\frac{2\pi}{ka}}\,e^{i\frac{\pi}{4}}\frac{1}{\sqrt{(N+m)(1-\zeta_{N+m}^2)}}\\ &\times\exp\left\{i\left[ka(N+m)\frac{1+\zeta_{N+m}^2}{2}-\frac{\pi}{2}(N+m)\right]\right\}\\ &\times\left\{1+0\left[\frac{1}{ka(1-\zeta_{N+m}^2)}\right]+0\left[\frac{1}{kal(1-\zeta_{N+m}^2)^2}\right]\right\} \end{aligned} \tag{2.19}$$

The first term in (2.19) is due to the use of formula (2.7),[3] whereas the second term specifies the error of the steepest-descent method. The latter will decrease with

3) This term can be diminshed if we retain in the asymptotic formula (2.7) not only the principal term, ζ_{N+m}

increasing $ka\ell$ ($\ell > 0$) if the following condition is imposed on the saddle points:

$$|\zeta_{N+m}| < 1 - \Delta |kal|^{-1/2+\delta}, \tag{2.20}$$

where Δ and $\delta < \frac{1}{2}$ are fixed positive constants. For obtaining asymptotic estimates indicating (for $|ka| \to \infty$) only the rate at which the error tends to zero, it suffices to regard the numerical coefficient Δ as a fixed quantity of the order of unity, i.e. $\Delta = 0(1)$. The actual value of the coefficient Δ that minimizes the error of calculation of the integral $I_N(\ell)$ and of the integral $H_N(\ell)$ (see page 103 below) will depend on the parameters of the problem and can be determined by numerical calculations only. The inequality (2.20) will hold for all $m = 0, 1, 2, \ldots$, if

$$N > \frac{l}{\pi}(1 - \Delta |kal|^{-1/2+\delta})^{-1} \sim \frac{l}{\pi}(1 + \Delta |kal|^{-1/2+\delta}). \tag{2.21}$$

Condition (2.21) coincides with condition (1.20) of separate occurrence of the wave U_N that has been obtained in the previous section on the basis of simple geometrical-optics considerations. It will be assumed that condition (2.21) holds also below.

Using the formula (1.17) for the propagation time T_{N+m} of a wave that arrives at the observation point after $N + m$ crossings of the waveguide axis, we can rewrite formula (2.19) as follows:

$$s_{N+m} = \sqrt{\frac{2\pi}{ka}}\, e^{i\frac{\pi}{4}} \frac{e^{i\left[T_{N+m}\omega - \frac{\pi}{2}(N+m)\right]}}{\sqrt{J_{N+m}}}$$

$$\times \left\{1 + 0\left[\frac{1}{\Delta (kal)^{\delta}}\left(\frac{l}{ka}\right)^{1/2}\right] + 0\left[\frac{1}{\Delta^2 (kal)^{2\delta}}\right]\right\}, \tag{2.22}$$

where

$$J_{N+m} = (N + m)(1 - \zeta^2_{N+m}) = (N + m) \sin^2 \psi_{N+m}$$

is the geometrical deviation for the corresponding waves; the factor $\sqrt{2\pi/ka}\, e^{i\pi/4}$ accounts for the action of the source, and the term $\pi(N + m)/2$ in the exponent specifies the phase delay due to $(N + m)$ -fold tangency of the ray and the caustics.

Thus for $|ka| \gg 1$ and values of $\ell \leqslant$ const $|ka|$, where const $\sim$ 1, the integrals s_{N+m} will describe waves U_{N+m} that arrive at the observation point after $(N + m)$ -fold crossing of the waveguide axis. If $\ell \gg ka$, we must write the condition (2.20) differently, i.e. $\zeta_{N+m} < 1 - \Delta|ka|^{-1+2\delta}$ (cf. (1.22)). However, this case will not be considered by us. For values of $N + m$ that satisfy (2.21) it follows from (2.22) that the waves U_{N+m} will have very distinct geometrical characteristics. Let us note that formula (2.22) can be obtained by the ray method. If ζ_{N+m} does not satisfy the inequality (2.20), the error of the steepest-descent method will be unsatisfactory and (2.22) will not apply. This means that such waves U_{N+m} have lost their geometrical characteristics. However, since we assume (2.21) to be valid, it follows that integrals s_{N+m} with such numbers $N + m$ are not contained in the sum (2.16). The totality of waves for which the number of crossings of the waveguide axis does not satisfy the inequality (2.21) is described by the integral $H_N(\ell)$. Thus the integral H_N will represent a wave perturbation propagating along the waveguide axis. On the other hand, it is evident that the integral S_{N+m} (2.18) describes waves that cross the waveguide axis $N + M$ times or more. The total wave field corresponding to such waves can be calculated by tabulating the integral S_{N+M}. The tabulation is facilitated by the fact that with an appropriate choice of the integration contour, the integrand of S_{N+M} will have a sufficiently sharp maximum. The wave fields of the individual waves can be obtained by separating from the integral S_{N+M} further terms of the type of s_{N+m} integrals.

4. Now let us calculate the wave field of an "axial wave," i.e., of a wave perturbation that propagates along the waveguide axis and is described by the integral $H_N(\ell)$.

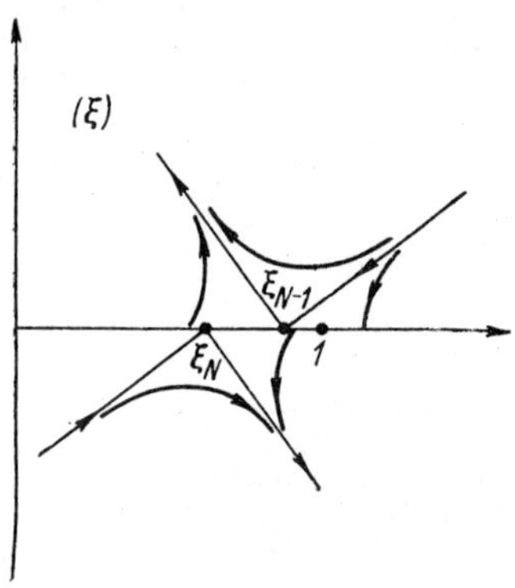

Figure 5

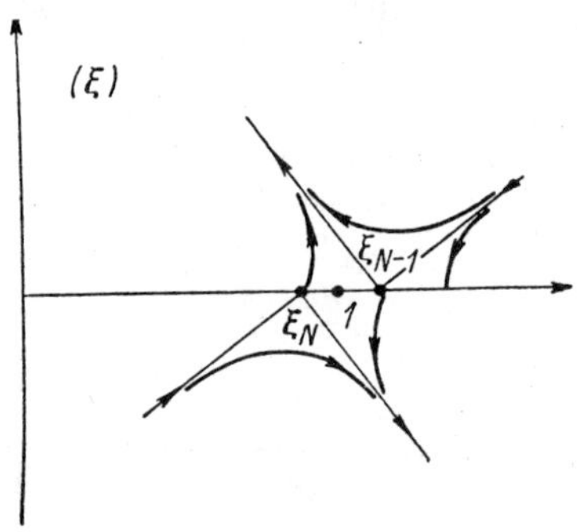

Figure 6

In the first quadrant to the right of the pole line (1.11) and in the fourth quadrant apart from a neighborhood of the pole line the integrand of the integral H_N will behave in the case of sufficiently large values of $|\mathrm{Im}\zeta|$ as follows:

$$\Phi_N(\zeta) = \begin{cases} e^{i\pi t(N-1)+ikal\zeta}, & \mathrm{Im}\,\zeta > 0, \\ e^{i\pi tN+ikal\zeta}, & \mathrm{Im}\,\zeta < 0. \end{cases}$$

The lines of fastest descent of the function $\Phi_N(\zeta)$ are plotted in Figures 5 and 6. Figures 5 applies to the case $\zeta_{N-1} = 1/\pi(N-1) < 1$, and Figure 6 to the case $\zeta_{N-1} = \ell/\pi(N-1) > 1$.

In the second case the function $\Phi_N(\zeta)$ decreases on the integration contour λ_1 on both sides of the point $\zeta = 1$, whereas in the first case it increases at first in the upper half-plane of the function $\Phi_N(\zeta)$ and only after some distance from the point $\zeta = 1$ does it begin to decrease. In both cases the integrand of the integral $H_N(l)$ (see (2.14)) has on the contour λ_1 a maximum in a neighborhood of the point $\zeta = 1$. This maximum will be the sharper, the larger the value of the large parameter $|ka|$. This circumstance facilitates the tabulation of the integral $H_N(l)$. For estimating the rate of convergence of $H_N(l)$ in a neighborhood of the point $\zeta = 1$, and also for giving the integral $H_N(l)$ a form most convenient for tabulation, we shall go over from the variable ζ to the variable $\tau = ka(1 - \zeta^2)/2$.

In the variable τ the integral $H_N(l)$ assumes the form

$$H_N(l) = -i\sqrt{\frac{\pi}{2ka}}\, e^{-i\frac{\pi}{2}N} \int\limits_L 2^{-\tau} \frac{\Gamma\left(\tau + \frac{1}{2}\right)}{\Gamma^2\left(\frac{\tau}{2} + \frac{3}{4}\right)} \times \frac{e^{i\pi\tau N + ikal\sqrt{1-\frac{2}{ka}\tau}}}{1 + ie^{i\pi\tau}} \frac{d\tau}{\sqrt{1 - \frac{2}{ka}\tau}}. \qquad (2.23)$$

In the τ-plane we made a cut $\operatorname{Re}\tau > ka/2$, and fixed the branch of the root $\sqrt{1 - 2\tau/ka}$ by the condition $\sqrt{1 - 2\tau/ka} = 1$ for $\tau = 0$.

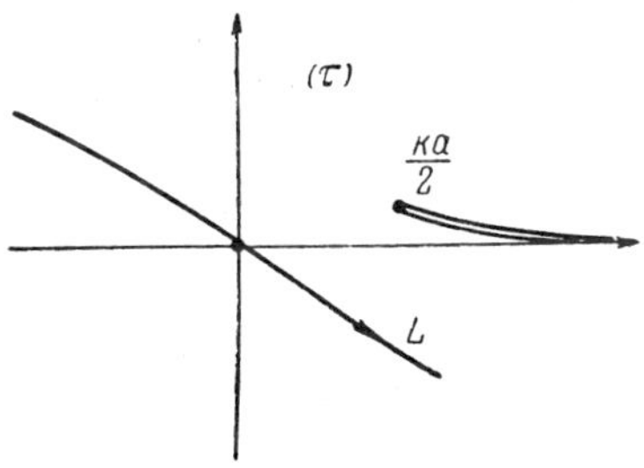

Figure 7.

The integration contour L that passes through the point $\tau = 0$ is shown in Figure 7. The principal contribution to the integral in (2.23) is made by integration over a small section of the contour L in a neighborhood of the point $\tau = 0$. Let us estimate the magnitude of this section.

We shall assume that in a neighborhood of the point $\tau = 0$ the integration contour L coincides with the straight line $\tau = 3\pi/4, -\pi/4$, and that for the principal section we have

$$\left|\frac{\tau}{ka}\right| \ll 1. \qquad (2.24)$$

By virtue of the inequality (2.24) we have

$$\sqrt{1-\frac{2}{ka}\tau}=1-\frac{1}{ka}\tau-\frac{1}{2}\left(\frac{1}{ka}\right)^2\tau^2+0\left[\left(\frac{\tau}{ka}\right)^3\right].$$

For values of $|\tau|$ of the order of a few units, it is possible to estimate the modulus of the integrand function on the upper part of the integration contour by the formula

$$\exp\left\{-\frac{1}{\sqrt{2}}\left(\pi N-l+\frac{l}{\sqrt{2}}\frac{|\tau|}{|ka|}\right)|\tau|+0\left[\frac{l\,|\tau|^3}{|ka|^2}\right]\right\}, \tag{2.25}$$

and on the lower part by the formula

$$\exp\left\{\left[\frac{1}{\sqrt{2}}\left(\pi N-\pi-l-\frac{l}{\sqrt{2}}\frac{|\tau|}{|ka|}\right)|\tau|+0\left[\frac{l\,|\tau|^3}{|ka|^2}\right]\right\}. \tag{2.26}$$

The function (2.26) begins to decrease only for

$$|\tau|>\rho_1=\frac{\sqrt{2}}{l}(\pi N-\pi-l)\,|ka|.$$

For integer N we have the inequality (2.21), which bounds the choice of N from below. The value of ρ_1 will be the smaller, the smaller the number N. Therefore we must take N as the smallest number satisfying (2.21). Hence

$$\frac{l}{\pi}(1+\Delta\,|kal\,|^{-1/2+\delta})<N\leqslant\frac{l}{\pi}(1+\Delta\,(kal)^{-1/2+\delta})+1. \tag{2.27}$$

If N satisfies (2.27), then for

$$|\tau|=\rho_0=2\Delta\left|\frac{ka}{l}\right|^{1/2}|kal\,|^{\delta}>\rho_1$$

on the contour L we have the bound

$$\left|\frac{e^{i(\pi N-l)\tau-i\frac{1}{2}l\frac{1}{ka}\tau^2}}{1+ie^{i\pi\tau}}\right|<e^{-C|kal\,|^{2\delta}}\leqslant 1, \tag{2.28}$$

$$C=(2-\sqrt{2})\,\Delta^2>0.$$

The bound (2.28) shows that the principal contribution to the integral (2.23) is made by integration over a section of the contour L expressed by

$$|\tau|<\rho_0. \tag{2.29}$$

On the principal integration path (2.29) the correction term in (2.25) and (2.26) will not exceed a quantity of the order of $8\Delta^3|ka\ell|^{-1/2+3\delta}$ Hence if the number N satisfies the condition (2.27), we can write the integral H_N in the form

$$H_N(l)=i\sqrt{\frac{\pi}{2ka}}\,e^{i\left(kal-\frac{\pi}{2}N\right)}\int\limits_{\rho_0 e^{-i\frac{\pi}{4}}}^{\rho_0 e^{i\frac{3\pi}{4}}} 2^{-\tau}\frac{\Gamma\left(\tau+\frac{1}{2}\right)}{\Gamma^2\left(\frac{\tau}{2}+\frac{3}{4}\right)}$$

$$\times\frac{e^{i\left[(\pi N-l)\tau-\frac{1}{2}\frac{l}{ka}\tau^2\right]}}{1+ie^{i\pi\tau}}\,d\tau\left\{1+0\left[\frac{\Delta^3}{(kal)^{1/2-3\delta}}\right]\right\}. \qquad (2.30)$$

It is natural to select the quantity δ in such a way as to minimize the error in the evaluation of the integrals s_{N+m} and H_N. If $\ell\sim 1$, the error of the formula (2.22) will be specified by the quantity $0(1/\Delta^2(ka\ell)^{2\delta})$. If $\delta=1/10$, then this error will coincide with the error of formula (2.30), being of the order of $|1/ka\ell|^{1/5}$. If $\ell\sim|ka|$, then the error of the formula (2.22) will be specified by the quantity $0(1/(ka\ell)^{\delta})$; in this case δ must be set equal to 1/8. But the error of (2.22) and (2.30) will be equal also in this case to $0[(1/ka\ell)^{1/5}]$, if in the asymptotic formula (2.7) we take into account the first correction term and set (as before) $\delta=1/10$. Without changing the magnitude of the error, it is possible to regard the integration limits in (2.30) as infinite, so that we finally obtain

$$H_N(l)=\sqrt{\frac{\pi}{2ka}}\,e^{i\left[kal-\frac{\pi}{2}(N-1)\right]}V_N\left(l,\frac{l}{ka}\right)\left\{1+0\left[\frac{\Delta^3}{(kal)^{1/5}}\right]\right\}, \qquad (2.31)$$

where

$$V_N(\xi,\eta)=\int\limits_{\infty e^{-i\frac{\pi}{4}}}^{\infty e^{i\frac{3\pi}{4}}} 2^{-\tau}\frac{\Gamma\left(\tau+\frac{1}{2}\right)}{\Gamma^2\left(\frac{\tau}{2}+\frac{3}{4}\right)}\frac{e^{i\left[(\pi N-\xi)\tau-\frac{1}{2}\eta\tau^2\right]}}{1+ie^{i\pi\tau}}\,d\tau, \qquad (2.32)$$

$$\frac{\xi}{\pi}\left(1+\Delta\left|\frac{\xi^2}{\eta}\right|^{-1/2+\delta}\right)<N\leqslant\frac{\xi}{\pi}\left(1+\Delta\left|\frac{\xi^2}{\eta}\right|^{-1/2+\delta}\right)+1,$$

is a special function that must be tabulated.

Thus the dispersion of an axial wave and the variation of its intensity with the distance between the observation point and the source are fully determined by the properties of the function $V_N(\xi,\eta)$. The number N in (2.31) varies in a jump-like manner with increasing ℓ. Let ℓ_0 be such that

$$\frac{l_0}{\pi}\left[1+\Delta|kal_0|^{-1/2+\delta}\right]$$

is an integer N_0. Hence for $\ell = \ell_0 - \varepsilon$, $\varepsilon > 0$, the number N in (2.16) and (2.31) will be equal to N_0, whereas for $\ell = \ell_0$ we have $N = N_0 + 1$ and in the expansion (2.16) it becomes smaller by one term, namely s_{N_0}. With the aid of the relation

$$\frac{e^{i\pi(N+1)\tau}}{1+ie^{i\pi\tau}} = i\frac{e^{i\pi N\tau}}{1+ie^{i\pi\tau}} - ie^{i\pi N\tau},$$

it is easy to see that to within correction terms we have

$$H_{N_0+1}(l) - H_{N_0}(l) = s_{N_0}.$$

Hence for values of ℓ such that

$$\frac{l}{\pi}[1+\Delta|kal|^{-1/2+\delta}] = N,$$

where N is an integer, the wave with number N, i.e., the wave that crosses N times the waveguide axis, will lose its geometrical characteristics, being absorbed by the axial wave.

§3. Axial waves in a symmetrical waveguide of arbitrary cross-section

In this section we shall extend the results obtained in the study of the reference problem to a more general case.

We shall assume that the velocity of wave propagation depends only on one coordinate y and that the relative refractive index $n(y) = c_0/c(y)$ is described by a symmetrical function that can be expanded in the case of small y in the form

$$n(y) = 1 - n_2 y^2 + n_4 y^4 + \dots, \quad n_2 > 0, \tag{3.1}$$

tending to a constant n_∞ $(0 < n_\infty < 1)$ when $|y| \to \infty$.

As before, we shall assume that the source and the observation point are located on the waveguide axis at points with coordinates $(0, 0)$ and $(x, 0)$.

In the first part of this section we shall analyze the behavior of rays in such a problem, and reach the conclusion that there exist two essentially different types of waveguides. The second part of this section is devoted to a generalization of formula (2.31) for both these types of waveguides.

1. Let ψ be the angle of exit of a ray from the source, i.e. the angle of inclination of the tangent to the ray at the source, and let $\cos\psi = \zeta$. The equation of a ray leaving the source at an angle ψ can be easily obtained from Euler's equation for a geometrical-optics functional:

$$T = \frac{1}{c_0}\int n(y)\,ds. \tag{3.2}$$

For small x the ray equation will be

$$x = \zeta \int_0^y \frac{d\tau}{\sqrt{n^2(\tau) - \zeta^2}}. \tag{3.3}$$

Equation (3.3) holds for $0 \leqslant y < y_m(\zeta)$, where $y_m(\zeta)$ is the positive root of the equation $n^2(y) - \zeta^2 = 0$, and hence for values of x belonging to the interval $0 \leqslant x \leqslant x_{1/2}$ we have

$$x_{1/2} = \zeta \int_0^{y_m} \frac{d\tau}{\sqrt{n^2(\tau) - \zeta^2}}.$$

For writing the ray equation for $x > x_{1/2}$, we shall divide the positive axis into adjoining intervals

$$x_{j-1/2} \equiv (2j-1)\, x_{1/2} \leqslant x \leqslant (2j+1)\, x_{1/2} \equiv x_{j+1/2}, \quad j = 1, 2, 3, \ldots \tag{3.4}$$

In each interval (3.4) the ray equation has the form

$$x = x_j + \zeta \int_0^y \frac{d\tau}{\sqrt{n^2(\tau) - \zeta^2}}, \tag{3.5}$$

where x_j is specified by the formula

$$x_j \equiv 2j x_{1/2} = 2j\zeta \int_0^{y_m} \frac{d\tau}{\sqrt{n^2(\tau) - \zeta^2}}$$

representing the coordinate of the point at which a ray emerging from the source at an angle ψ crosses the waveguide axis for the jth time. For ascertaining the behavior of the function $x_j(\zeta) = x_j\,(\cos\psi)$ for small ψ, we shall expand $x_j\,(\cos\psi)$ in a power series in ψ:

$$x_j = \pi a j \left[1 - \frac{b}{2}\psi^2 + O(\psi^4)\right], \tag{3.6}$$

where $a = 1/\sqrt{2n_2}$ is the characteristic dimension of the waveguide, and

$$b = \frac{5}{8} - \frac{3}{4}\frac{n_4}{n_2^2} \tag{3.7}$$

is a dimensionless quantity. It follows from (3.6) that $\lim_{\psi\to 0} x_j = \pi a j$; that is, when the angle of exit of the ray from the source tends to zero, the points of intersection x_j of the ray with the waveguide axis will tend to limiting values that are at equal distances from the source and from each other. Let us note that if

$$n(y) = 1 - n_1 |y| + 0(y^2), \qquad n_1 > 0,$$

i.e. if the derivative of the refractive index at the waveguide axis undergoes a jump (or, in the case of a surface waveguide, when we have the "whispering gallery" effect), then the points of intersection of the ray with the waveguide axis (or with the boundary of the region) will unboundedly approach the source with decreasing angle of exit of the ray from the source. This causes major differences in the propagation of waves in a smooth waveguide as compared to a waveguide that has a break in the refractive index.

Figure 8

If $b > 0$ in (3.6), then for $\psi \to 0$ the points x_j will approach the limiting values $\pi a j$ from the side of smaller values, whereas for $b < 0$ they will approach them from the side of larger values. In the first case, i.e. for $b > 0$, we shall say that the waveguide is strong, and in the second case (i.e. for $b < 0$) that it is weak. In Figure 8 we represent by solid lines the ray propagation in a strong waveguide, and by dashed lines the ray propagation in a weak waveguide. Strong waveguides can be also called sharply outlined waveguides, since in such waveguides the rays undergo stronger bending.

Let us calculate the propagation time T_j of a perturbation along a ray that emerges from the source at an angle ψ and arrives at the waveguide axis at a point x_j after j-fold crossing of the waveguide axis. Using the ray equations (3.3) and (3.5), we obtain for the quantity T_j specified by the integral (3.2) the formula

$$T_j = \frac{1}{c_0}\left[\zeta x_j + j2\int_0^{y_m} \sqrt{n^2(\tau) - \zeta^2}\, d\tau\right]$$

The time T_0 of propagation of a perturbation from the source to the observation point x_j along the waveguide axis is evidently equal to

$$T_0 = \frac{x_j}{c_0} = \frac{1}{c_0} 2j\zeta \int_0^{y_m} \frac{d\tau}{\sqrt{n^2(\tau) - \zeta^2}}$$

It is easy to find the value of the difference $T_j - T_0$ for small angles of exit ψ of a ray from the source. By expanding the integrals occurring in $T_j - T_0$ in powers of ψ, we obtain

$$T_j - T_0 = j\frac{\pi a}{c_0}\frac{b}{8}\psi^4 + O(\psi^6). \tag{3.8}$$

It follows from (3.8) that in a strong waveguide ($b > 0$) the first to arrive is the perturbation that travels along the waveguide axis (the axial wave), and only after it do we get the waves that cross a certain (different) number of times the waveguide axis. In contrast, in a weak waveguide the axial wave will be the last to arrive.

Natural waveguides (such as a layer of reduced velocity of propagation of sound in the ocean, or a layer of reduced velocity of propagation of electromagnetic waves in the ionosphere) are usually weak waveguides. For the reference problem considered in §§1 and 2 we have $b > 0$, and hence in this problem the waveguide will be strong. With the aid of (3.6) it is possible, with a fixed coordinate x of the observation point and a given number j of crossings of the waveguide axis by the ray, to determine the angle of exit $\psi_j(x)$ of the ray from the source. By introducing the dimensionless coordinate $\ell = x/a$ of the observation point, we obtain

$$\psi_j(l) = \pm\sqrt{\frac{2}{b}\left(1-\frac{l}{\pi j}\right)}\cdot\left\{1+O\left[1-\frac{l}{\pi j}\right]\right\}. \tag{3.9}$$

It follows from the form of the remainder that formula (3.9) holds for small values of the difference $1 - \ell/\pi j$, i.e. for small j for distances ℓ such that $|\ell - \pi j| \ll 1$, or for large j for distances $|\ell - \pi j| \ll C$, where the quantity $C \ll j$ is of the order of unity. Rays emerging from the source at small angles must satisfy the inequalities

$$\begin{aligned} j &\geqslant \ell/\pi \quad \text{for} \quad b > 0 \text{ (for strong waveguides)} \\ j &\leqslant \ell/\pi \quad \text{for} \quad b < 0 \text{ (for weak waveguides)}\,. \end{aligned} \tag{3.10}$$

By substituting the value of the angle ψ (3.9) into (3.8), we obtain the value of the difference $T_j - T_0$ as a function of the coordinate ℓ of the observation point and of the number j of crossings of the waveguide axis by the ray:

$$T_j - T_0 = \frac{\pi a}{c_0}\frac{1}{2b}j\left(1-\frac{l}{\pi j}\right)^2 + O\left[j\left(1-\frac{l}{\pi j}\right)^3\right]. \tag{3.11}$$

As in §1, we shall assume that a wave that crosses the waveguide axis j times can be observed independently of the wave U_0 if the inequality

$$|T_j - T_0| > q\,(kal)^{2\delta}\,T, \quad q = O(1),\ 0 < \delta \ll 1, \tag{3.12}$$

holds; here $T = 2\pi/\omega = 2\pi\tau_0/ka$, $k = \omega/c_0$, and $\tau_0 = a/c_0$ is the characteristic time of the waveguide. By substituting into (3.12) the approximate value of the difference $T_j - T_0$ (3.11) and neglecting small (for $ka \gg 1$) terms, we obtain

$$\begin{aligned} j &> \frac{l}{\pi}\left[1 + \Delta b^{1/2}(kal)^{-1/2+\delta}\right] \quad \text{for} \quad b > 0, \\ j &< \frac{l}{\pi}\left[1 - \Delta |b|^{1/2}(kal)^{-1/2+\delta}\right] \quad \text{for} \quad b < 0, \end{aligned} \tag{3.13}$$

where $\Delta = 2\sqrt{\pi q} = 0(1)$. The waves for which the number of crossings j satisfies (3.10) but not (3.13), i.e.

$$\begin{aligned} &\frac{l}{\pi} < j \leqslant \frac{l}{\pi}[1 + \Delta b^{1/2}(kal)^{-1/2+\delta}] \quad \text{for } b > 0, \\ &\frac{l}{\pi}[1 - \Delta |b|^{1/2}(kal)^{-1/2+\delta}] \leqslant j < \frac{l}{\pi} \quad \text{for } b < 0, \end{aligned} \tag{3.14}$$

will interfere with the wave propagating directly along the waveguide axis, and also with each other. As a result there appears a wave field of an axial wave of a fairly complicated structure.

In the next subsection we shall obtain formulas describing an axial wave in the case of a strong waveguide and in the case of a weak waveguide. We shall proceed from formula (2.30) for an axial wave in the reference problem.

2. The mathematical problem of wave propagation in a symmetrical waveguide can be formulated in the same way as in the case of the reference problem; it involves the solving of the equation

$$\frac{\partial^2 u}{\partial x^2} + \frac{\partial^2 u}{\partial y^2} + \frac{\omega^2}{c_0^2} n^2(y)\, u = 0 \tag{3.15}$$

under the conditions (1.4) and (1.5).

An asymptotic expression for the solutions of the homogeneous equation (3.15) for $\omega \to \infty$ in the vicinity of the waveguide axis $y = 0$ can be obtained in the form of a product of an exponential function and a parabolic-cylinder function

$$\begin{aligned} u_q(x, \nu) = \exp\left\{ i \sum_{n=-2}^{M-1} \alpha_n(x, \nu)\, \omega^{-n/2} + 0(\omega^{-M}) \right\} \\ \times D_q\left\{ \sqrt{2} \sum_{n=0}^{M-1} \beta_n(x, \nu)\, \omega^{-n/2} + 0(\omega^{-M}) \right\}, \end{aligned} \tag{3.16}$$

whose arguments are sections of formal infinite series in negative powers of $\omega^{1/2}$. The coefficients of these series $\alpha_n(x,\nu)$ and $\beta_n(x,\nu)$, which are polynomials in $\nu = \omega^{1/2} y$, can be obtained from a recursive system of differential equations, resulting from the substitution of (3.16) into (3.15) [1, 3]. Let us note that the asymptotic expression for the solutions of equation (3.15) can be also constructed by separation of variables in equation (3.15) with subsequent use of asymptotic formulas for the solutions of an ordinary differential equation with two turning points [4, 5]. However this second method is less advantageous, since it does not enable us to obtain an extension of the final "axial wave" formulas to the case of bent symmetrical waveguides.

Omitting the calculations performed in [1], which are greatly simplified in the case of a function $n(y)$ that is even and independent of x, we present only the final result:

$$u_q(x, \nu) = \sqrt[4]{\frac{c_0}{a}} \exp\left\{i\left(\frac{\omega}{c_0}x - \left(q+\frac{1}{2}\right)\frac{x}{a} + \left[i\frac{b-1}{8}\frac{1}{a^2}\nu^2\right.\right.\right.$$
$$\left.\left.\left. -\frac{1}{2}\left(b\left(q+\frac{1}{2}\right)^2 + \frac{b-1}{4}\right)\frac{c_0}{a^2}x - i\frac{1}{2}\left(q+\frac{1}{2}\right)\frac{c_0}{a}\right]\frac{1}{\omega} + 0\left[\frac{1}{\omega^2}\right]\right)\right\}$$
$$\times D_q\left\{\sqrt{2}\left(\frac{\nu}{\sqrt{ac_0}} + \frac{b-1}{12}\left[\frac{1}{a^{5/2}c_0^{1/2}}\nu^3 + 3\left(q+\frac{1}{2}\right)\frac{c_0^{1/2}}{a^{3/2}}\nu\right]\frac{1}{\omega} + 0\left[\frac{1}{\omega^2}\right]\right)\right\}, \quad (3.17)$$

where $|\nu| <$ const, and the arbitrary elements contained in the solutions (3.16) are selected in such a way that for $n(y) = (1 - 2ny^2)^{1/2}$ the formula (3.17) goes over into an asymptotic expression for the solutions of the reference problem.

For describing an axial wave on the waveguide axis, i.e. for $\nu = 0$, we shall construct from the solutions $u_q(s, \nu)$, calculated for $\nu = 0$, a linear superposition H_N. This superposition will have a different form for a strong ($b > 0$) and a weak ($b < 0$) waveguide.

First we shall consider a waveguide of the same type as in the reference problem, i.e. a strong waveguide. The linear superposition H_N will be constructed by analogy with the integral (2.30) obtained in the study of the reference problem:

$$H_N(l) = \int_{\rho_0 e^{-i\pi/4}}^{\rho_0 e^{i3\pi/4}} C(\omega, \tau)\, u_{\tau-1/2}(al, 0)\, d\tau. \quad (3.18)$$

The weight function $C(\omega, \tau)$ and the quantity ρ_0 must be selected in such a way that: 1) At the ends of the integration interval the integrand will assume exponentially small values of the order of $\exp[-\Delta_1 \omega^\varepsilon]_{\Delta_1>0, \varepsilon>0}$;[4] 2) the integral H_N will describe a wave field that corresponds to a source of selected intensity; 3) the reciprocity principle will hold. The latter condition will be satisfied if we write

$$C(\omega, \tau) = \omega^{-1/2} p(\tau)\, u_{\tau-1/2}(0, 0),$$

where the factor $\omega^{-1/2}$ has been selected from considerations of dimensionality. The new, dimensionless weight function $p(\tau)$ must depend neither on the coordinates of the observation point, nor on the coordinates of the source. We shall assume the function $p(\tau)$ to be independent also of the frequency ω.[5] By introducing the function $C(\omega, \tau)$ and the values of $u_{\tau-1/2}(al, 0)$ and $u_{\tau-1/2}(0, 0)$ specified by formula (3.17) into the integral (3.18), we obtain

$$H_N(l) = \omega^{-1/2}\sqrt{\frac{c_0}{a}}\, e^{i\left(\frac{\omega}{c_0}a - \frac{b-1}{8}\frac{c_0}{a}\frac{1}{\omega}\right)l}$$
$$\times \int_{\rho_0 e^{-i\pi/4}}^{\rho_0 e^{i3\pi/1}} p(\tau)\,[D_{\tau-1/2}(0)]^2\, e^{-i\tau l - \left(i\frac{b}{2}l\tau^2 - \tau\right)\frac{c_0}{a}\cdot\frac{1}{\omega}}\left\{1 + 0\left(\frac{l\tau^3}{\omega^2}\right)\right\} d\tau. \quad (3.19)$$

---

[4] Under this condition the exponential asymptotic expression for the wave field will not depend on ρ_0.

[5] As we can show, this assumption is equivalent to the principle of locality, which asserts that the high-frequency asymptotic behavior of a wave field depends only on the properties of the medium in the vicinity of the ray.

If ℓ is large, the factor $e^{-i\tau\ell-i(b/2)\ell\tau^2(c_0/a)(1/\omega)}$ in the integrand will begin to decrease on the upper part of the integration contour only for sufficiently large values of τ. For the error of evaluation of the integral H_N to be independent of ℓ, it is necessary to select the weight function $p(\tau)$ differently for different values of ℓ.

Let the coordinate $\ell \lesssim ka$ be fixed. We shall select an integer N such that

$$\frac{l}{\pi}\left[1+\Delta b^{1/2}\left|\frac{a}{c_0}\omega l\right|^{-1/2+\delta}\right] < N \leqslant \frac{l}{\pi}\left[1+\Delta b^{1/2}\left|\frac{a}{c_0}\omega l\right|^{-1/2+\delta}\right]+1, \tag{3.20}$$

where $\Delta = 0(1)$ and $0 < \delta \ll 1$. By virtue of condition (3.13) the number N is equal to the minimum number of crossings of the waveguide axis by a ray such that the wave U_N corresponding to this ray will be governed by geometrical laws and can be observed independently of the axial wave. On the other hand, when we go over to the reference problem ($b \to 1$), the inequality (3.20) will go over into the inequality (2.27), for which the axial wave is described by the integral (2.30). The weight function $p(\tau)$, which does not depend on the parameters of the problem, and which is therefore the same for all the problems of this class, must be selected in such a way that for $b \to 1$ the integral (3.19) goes over into the integral (2.30). If we write

$$p(\tau)=\sqrt{\frac{\pi}{2}}\,e^{-i\frac{\pi}{2}(N-1)}2^{-\tau}\frac{\Gamma\left(\tau+\frac{1}{2}\right)}{\Gamma^2\left(\frac{\tau}{2}+\frac{3}{4}\right)}\frac{e^{i\pi N\tau}}{1+ie^{i\pi\tau}}\cdot[D_{\tau-1/2}(0)]^{-2},$$

the desired transition will take place.

Thus if the condition (3.20) holds, the axial wave will be described by the formula

$$H_N(l)=\sqrt{\frac{\pi}{2ka}}\exp\left\{i\left[kal-\frac{b-1}{8}\frac{l}{ka}-\frac{\pi}{2}(N-1)\right]\right\}$$

$$\times\int_{\rho_0e^{-i\pi/4}}^{\rho_0e^{i3\pi/4}} 2^{-\tau}\frac{\Gamma\left(\tau+\frac{1}{2}\right)}{\Gamma^2\left(\frac{\tau}{2}+\frac{3}{4}\right)}\frac{e^{i\left(\pi N-l-\frac{i}{ka}\right)\tau-i\frac{b}{2}\frac{l}{ka}\tau^2}}{1+ie^{i\pi\tau}}\left\{1+0\left[\frac{l\tau^3}{(ka)^2}\right]\right\}d\tau, \tag{3.21}$$

where (as before) $k = \omega/c_0$ is the wave number. The quantity ρ_0, which specifies the size of the integration contour, can be taken equal to

$$\rho_0=2\Delta b^{1/2}\left(\frac{ka}{l}\right)^{1/2}(kal)^{1/10}. \tag{3.22}$$

In the same way as in §2 we can check whether or not in the case of (3.22) the integrand assumes exponentially small values at the ends of the integration interval. Let us introduce the following function:

$$V_N(\xi, \eta) = \int_{\infty e^{-i\pi/4}}^{\infty e^{i3\pi/4}} 2^{-\tau} \frac{\Gamma\left(\tau + \frac{1}{2}\right)}{\Gamma^2\left(\frac{\tau}{2} + \frac{3}{4}\right)} \frac{e^{i\left[(\pi N - \xi)\tau - \frac{1}{2}\eta\tau^2\right]}}{1 + ie^{i\pi\tau}} d\tau. \tag{3.23}$$

With the aid of the function $V_N(\xi, \eta)$ it is possible to write an axial wave, within terms of the order of $(ka\ell)^{-1/5}$, as follows:

$$H_N(l) = \sqrt{\frac{\pi}{2ka}} e^{i\left[kal - \frac{b-1}{8}\frac{l}{ka} - \frac{\pi}{2}(N-1)\right]} V_N\left(l, b\frac{l}{ka}\right). \tag{3.24}$$

For $b = 1$ the formula (3.24) goes over into (2.31), which describes an axial wave in the reference problem. Let us note that for $b \neq 1$ we cannot drop the term $(b-1)\ell/8ka$ in the exponent, since for $\ell \sim ka$ it will be of the order of magnitude of unity.

Now let us consider the case of a weak waveguide ($b < 0$). We shall take the number N in such a way that $N - 1$ is the maximum number for which the second inequality (3.13) holds, i.e.

$$\frac{l}{\pi}[1 - \Delta|b|^{1/2}|kal|^{-1/2+\delta}] \leqslant N < \frac{l}{\pi}[1 - \Delta|b|^{1/2}|kal|^{-1/2+\delta}] + 1. \tag{3.25}$$

With such a choice of the number N, the wave U_{N-1} corresponding to a ray that crosses the waveguide axis $N - 1$ times will be the last wave that does not interfere with the wave propagating along the waveguide axis and can be observed independently of the latter.

With such a choice of the number N and $b < 0$, the phase function $i(\pi N - \ell)\tau - ib\ell\tau^2/2ka$ of the integral (3.21) will differ by its sign from the above considered case of a strong waveguide. For the modulus of the integrand function to assume (as before) its maximum value on the integration contour in a neighborhood of the origin, it is necessary in the case of a weak waveguide to select the integration contour in the form of a straight-line segment connecting the points $\rho_0 e^{-i3\pi/4}$ and $\rho_0 e^{i\pi/4}$, $\rho_0 = 2\Delta|b|^{1/2}(ka/\ell)^{1/2}(ka\ell)^{\delta}$. Thus in the case of a weak waveguide the axial wave will be described by the integral

$$H_N(l) = \sqrt{\frac{\pi}{2ka}} \exp\left\{i\left[kal - \frac{b-1}{8}\frac{l}{ka} - \frac{\pi}{2}(N-1)\right]\right.$$
$$\times \int_{\rho_0 e^{-i3\pi/4}}^{\rho_0 e^{i\pi/4}} 2^{-\tau} \frac{\Gamma\left(\tau + \frac{1}{2}\right)}{\Gamma^2\left(\frac{\tau}{2} + \frac{3}{4}\right)} \frac{e^{i\left(\pi N - l - \frac{i}{ka}\right)\tau - i\frac{b}{2}\frac{l}{ka}\tau^2}}{1 + ie^{i\pi\tau}} \left\{1 + O\left[\frac{l\tau^3}{(ka)^2}\right]\right\} d\tau.$$

By setting $\delta = 1/10$, we obtain for the axial wave $H_N(\ell)$ the formula

$$H_N(l) = \sqrt{\frac{\pi}{2ka}} \exp\left\{i\left[kal - \frac{b-1}{8}\frac{l}{ka} - \frac{\pi}{2}(N-1)\right]\right\}$$
$$\times W_N\left(l, b\frac{l}{ka}\right)\left\{1 + 0\left[\left(\frac{1}{kal}\right)^{1/5}\right]\right\}, \qquad (3.26)$$

where

$$W_N(\xi, \eta) = \int_{\infty e^{-i3\pi/4}}^{\infty e^{i\pi/4}} 2^{-\tau} \frac{\Gamma\left(\tau + \frac{1}{2}\right)}{\Gamma^2\left(\frac{\tau}{2} + \frac{3}{4}\right)} \cdot \frac{e^{i\left[(\pi N - \xi)\tau - \frac{1}{2}\eta\tau^2\right]}}{1 + ie^{i\pi\tau}} d\tau \qquad (3.27)$$

is a special function that must be tabulated. Let us note that in the first approximation (as follows from (3.24) and (3.26)) the axial wave will depend only on the coordinate ℓ of the observation point and the waveguide parameters a and b that are determined on the basis of the second and the fourth derivative of the refractive index for $y = 0$.

Let us examine the process of separation of a geometrical wave from the axial wave (3.26). Let

$$\frac{l_0}{\pi}\left[1 - \Delta |b|^{1/2} (kal_0)^{-1/2+\delta}\right] = N_0$$

be an integer. Then for $\ell = \ell_0$ we have $N = N_0$, whereas for $\ell = \ell_0 + \varepsilon$, $\varepsilon > 0$, the inequality (3.25) will be satisfied by the number $N = N_0 + 1$. Using the relation

$$\frac{e^{i\pi N_0\tau}}{1 + ie^{i\pi\tau}} = -i\frac{e^{i\pi(N_0+1)\tau}}{1 + ie^{i\pi\tau}} + e^{i\pi N_0\tau},$$

we can express $H_{N_0}(\ell)$ in the form

$$H_{N_0}(l) = H_{N_0+1}(l) + \sqrt{\frac{\pi}{2ka}} \exp\left\{i\left[kal - \frac{b-1}{8}\frac{l}{ka} - \frac{\pi}{2}(N_0 - 1)\right]\right\}$$
$$\times \int_{\infty e^{-i3\pi/4}}^{\infty e^{i\pi/4}} 2^{-\tau} \frac{\Gamma\left(\tau + \frac{1}{2}\right)}{\Gamma^2\left(\frac{\tau}{2} + \frac{3}{4}\right)} \cdot e^{i\left[(\pi N_0 - l)\tau - \frac{b}{2}\frac{l}{ka}\tau^2\right]} d\tau. \qquad (3.28)$$

Under the condition (3.25) the integral in the right-hand side of (3.28) can be evaluated by the method of steepest descent.

After performing the necessary calculations, we obtain

$$H_{N_0}(l) = e^{i\left[\frac{\pi}{4} - \frac{\pi}{2}(N_0-1)\right]} \sqrt{\frac{\pi}{2ka}} \sqrt{\frac{2}{|\pi N_0 - l|}}\, e^{i\left[kal + \frac{ka}{2}\frac{(\pi N_0 - l)^2}{bl} - \frac{b-1}{8}\frac{l}{ka}\right]} + H_{N_0+1}(l). \quad (3.29)$$

The first term in the right-hand side of (3.29) describes a wave U_{N_0} that has been separated from an axial wave and has geometrical characteristics. The second term in (3.29) represents a new axial wave. If the wave U_{N_0} is determined by the methods of geometrical acoustics, we would obtain a formula that coincides with the first term of (3.29). This can be regarded as proof that the formula (3.26) indeed describes an axial wave in the case of a weak waveguide.

Let us note that if the refractive index n depends on a single coordinate y, we could obtain the formulas (3.24) and (3.26) for an axial wave by the same methods as were used in §2 for the reference problem. The advantage of the above heuristic derivation of (3.24) and (3.26) consists in the fact that this derivation can be extended to problems with variables that cannot be separated.

In conclusion the author expresses his gratitude to B. Ja. Gel′činskiĭ and D.K. Ozerov for useful discussions of various problems of wave propagation in waveguides.

BIBLIOGRAPHY

1. V.S. Buldyrev, *Asymptotic behavior of solutions of wave equations that are concentrated near the axis of a two-dimensional waveguide in an inhomogeneous medium*, Problemy Mat. Fiz., no. 3, Izdat. Leningrad. Univ., Leningrad, 1968, pp. 5–30; English transl., Topics in Math. Phys., no. 3, Consultants Bureau, New York, 1969, pp. 1–23.
2. L.M. Brehovskiĭ, *Waves in layered media*, Izdat. Akad. Nauk SSSR, Moscow, 1957; English transl., Academic Press, New York, 1960. MR **20** #1476; **22** #3243.
3. V.F. Lazutkin, *Construction of an asymptotic series of eigenfunctions of the "bouncing ball" type*, Trudy Mat. Inst. Steklov. **95** (1968), 106–118 = Proc. Steklov Inst. Math. **95** (1968), 125–140. MR **42** s 2194.
4. S. Ju. Slavjanov, *Asymptotic behavior of singular Sturm-Liouville problems for large parameter in the case of (nearby) transition points*, Differencial′nye Uravnenija **5** (1969), 313–325. (Russian) MR **39** #7227.
5. V.S. Buldyrev and S. Ju. Slavjanov, *Uniform asymptotic expansions for solutions of an equation of Schrödinger type with two transition points*. I, Vestnik Leningrad. Univ. **23** (1968), no. 22, 70–84. (Russian) MR **39** #3775.

Translated by:
L. Ebner

CONSTRUCTION OF SOLUTIONS CONCENTRATED CLOSE TO RAYS FOR THE EQUATIONS OF ELASTICITY THEORY IN AN INHOMOGENEOUS ISOTROPIC SPACE

N. JA. KIRPIČNIKOVA

UDC 517 944/947

Abstract. The paper offers formal solutions for the dynamic equations of elasticity theory with increasing frequency, these solutions being concentrated in the neighborhood of an isolated ray of longitudinal or transverse waves. These solutions can be obtained by using the procedures developed by V.M. Babič in his paper entitled "Eigenfunctions concentrated in the neighborhood of a closed geodesic" (MR 39 #3755). The resultant expressions correspond to highly directional propagation of elastic waves.

Bibliography: 2 items.

We consider an inhomogeneous isotropic elastic space and we introduce an arbitrary curvilinear coordinate system (q^1, q^2, q^3). The problem of wave propagation in this medium may be formulated mathematically as the problem of integrating the equations of elasticity theory:

$$\left.\begin{aligned} \sigma_{ij}\sqrt{g}\,\frac{\partial G^{ij}}{\partial q^r}+2\frac{\partial}{\partial q^i}(\sigma_{rj}G^{ij}\sqrt{g})-2\rho G_{ir}\sqrt{g}\,\frac{\partial^2\varphi^{(i)}}{\partial t^2}=0,\\ i,\ j,\ r=1,\ 2,\ 3, \end{aligned}\right\} \tag{1}$$

where we assume from now on that each letter index which is encountered in a product twice is a summation index over all possible values of this index.

In (1) the quantity $\bar{u} = (\varphi^{(1)}, \varphi^{(2)}, \varphi^{(3)})$ denotes the displacement vector of the elastic medium, t is the time, and G_{ij} are the coefficients of the metric tensor in the curvilinear coordinates (q^1, q^2, q^3), so that an element of length $d\sigma$ is equal to

$$d\sigma=\sqrt{G_{ij}dq^i dq^j}, \tag{2}$$

and the matrix $\|G_{ij}\|$ and the quantity g are specified by the formulas

$$\|G^{ij}\|=\|G_{ij}\|^{-1}, \quad g=\det\|G_{ij}\|. \tag{3}$$

The components σ_{ij} of the stress tensor are subject to Hooke's law for the medium under consideration:

AMS 1970 *subject classifications*. Primary 73D99; Secondary 35B40.

$$\sigma_{ij}=\frac{\lambda}{\sqrt{g}}G_{kj}\frac{\partial}{\partial q^r}(\sqrt{g}\,\varphi^{(r)})\delta_i^k$$

$$+\mu\left[\frac{\partial G_{ij}}{\partial q^r}\varphi^{(r)}+G_{ri}\frac{\partial\varphi^{(r)}}{\partial q^j}+G_{rj}\frac{\partial\varphi^{(r)}}{\partial q^i}\right],\quad r,\ i,\ j,\ k=1,\ 2,\ 3, \tag{4}$$

where δ_i^k is the Kronecker delta; $\lambda=\lambda(q^1,q^2,q^3)$ and $\mu=\mu(q^1,q^2,q^3)$ are the Lamé parameters; $\rho=\rho(q^1,q^2,q^3)$ is the density of the medium. The functions λ, μ and ρ are assumed to be differentiable a sufficient number of times.

The transverse velocity $b(q^1,q^2,q^3)$ and the longitudinal velocity $a(q^1,q^2,q^3)$ of propagation of elastic waves are defined by

$$b^2=\frac{\mu}{\rho},\quad a^2=\frac{\lambda+2\mu}{\rho}. \tag{5}$$

We denote by L_b the rays which are extrema of the integral

$$\int\frac{d\sigma}{b}, \tag{6}$$

and by L_a the extremal rays of the integral

$$\int\frac{d\sigma}{a}. \tag{7}$$

In this article we will construct asymptotic solutions for (1) which are concentrated near the rays L_b and L_a. The method used is that proposed by V.M. Babič in [1].

Particular attention is paid to the asymptotic behavior of solutions concentrated near the rays L_b. The changes which should be made in converting to L_a rays will be noted specially.

§1. All our subsequent constructions will be made in a special curvilinear coordinate system [2], which is introduced as follows.

We denote by s the arc length (taken with a particular sign) of the ray L between an arbitrary point M_s of the ray and some fixed point on it. For all points M_s on L which are on one side of the fixed point we choose one sign, and for the remaining points we take the opposite one, so that as s varies from $-\infty$ to $+\infty$ all the points of L will be obtained. Assume, furthermore, that $\bar{r}(s)$ is the radius vector of the point M_s of L, and that $\bar{n}(s)$ and $\bar{B}(s)$ are respectively the principal normal and the binormal of L at M_s. Then the radius vector $\bar{r}(M)$ of an arbitrary point M of the space may be written as

$$\bar{r}(M)=\bar{r}(s)+\eta\bar{e}_\eta+\zeta\bar{e}_\zeta, \tag{8}$$

where

$$\left.\begin{aligned}\bar{e}_\eta&=\bar{n}(s)\cos\vartheta-\bar{B}(s)\sin\vartheta,\\ \bar{e}_\zeta&=\bar{n}(s)\sin\vartheta+\bar{B}(s)\cos\vartheta.\end{aligned}\right\} \tag{9}$$

The quantities (s, η, ζ) define an orthogonal curvilinear coordinate system if the function $\theta = \theta(s)$ is selected such that

$$\vartheta(s) = -\int_0^s \varkappa(s)\, ds + \vartheta(0), \tag{10}$$

where $\kappa(s)$ is the torsion of L at M_s.

In our curvilinear coordinate system the ray L is specified by the equations

$$\eta = 0, \quad \zeta = 0. \tag{11}$$

We denote by $k(s)$ the curvature of L at $(s, 0, 0)$; then, using the Frenet formulas, we obtain

$$\frac{d\bar{e}_\eta}{ds} = -k(s)\frac{d\bar{r}(s)}{ds}\cos\vartheta, \quad \frac{d\bar{e}_\zeta}{ds} = -k(s)\frac{d\bar{r}(s)}{ds}\sin\vartheta. \tag{12}$$

Now we will have the following expressions for the components G_{ij} of the metric tensor:

$$\left.\begin{aligned} &G_{ss} = [1 - \eta k(s)\cos\vartheta - \zeta k(s)\sin\vartheta]^2, \\ &G_{\eta\eta} = 1, \quad G_{\zeta\zeta} = 1, \\ &G_{s\eta} = 0, \quad G_{s\zeta} = 0, \quad G_{\eta\zeta} = 0. \end{aligned}\right\} \tag{13}$$

The functional (6) assumes the form

$$\int \frac{\sqrt{G_{ss}ds^2 + d\eta^2 + d\zeta^2}}{b(s, \eta, \zeta)}. \tag{14}$$

We take as L any extremum of the functional (14), i.e. the ray L_b. Then the Euler equations must hold on this ray:

$$\left.\begin{aligned} &k(s)\cos\vartheta + \left(\frac{1}{b}\frac{\partial b}{\partial \eta}\right)\Big|_{\eta=0,\,\zeta=0} = 0, \\ &k(s)\sin\vartheta + \left(\frac{1}{b}\frac{\partial b}{\partial \zeta}\right)\Big|_{\eta=0,\,\zeta=0} = 0. \end{aligned}\right\} \tag{15}$$

Consider the family of extrema of (14) which are sufficiently close to L_b that the quantities η, ζ, $d\eta/ds$ and $d\zeta/ds$ corresponding to the extrema of this family are fairly small. Taking into account only the linear terms in these small quantities in the Euler equations, we obtain

$$\frac{D^2\bar{R}}{ds^2} - \frac{1}{b}\frac{\partial b}{\partial s}\frac{D\bar{R}}{ds} + K \cdot \bar{R} = 0, \tag{16}$$

where

$$\bar{R} = \eta(s)\,\bar{e}_{\eta} + \zeta(s)\,\bar{e}_{\zeta}, \tag{17}$$

$$K = K(s) = \frac{1}{b}\begin{Vmatrix} b''_{\eta\eta} & b''_{\eta\zeta} \\ b''_{\zeta\eta} & b''_{\zeta\zeta} \end{Vmatrix}, \tag{18}$$

while the matrix K is symmetrical and real, and

$$\eta = \eta(s), \quad \zeta = \zeta(s) \tag{19}$$

are the parametric equations of the extremum of the family under consideration.

The symbol D denotes differentiation with respect to the variable s for which the unit vectors $\bar{e}_\eta$ and $\bar{e}_\zeta$ are fixed and are not differentiated.

We should note that the coefficients of (16) and the matrix K are calculated on the ray L_b.

§2. To find the high-frequency asymptotic behavior of the solution of (1) which is concentrated along L_b, it turns out to be convenient to use the parabolic-equation method (see [1]). The parabolic equation can be solved explicitly, to within the solution of a system of ordinary differential equations of the form (16).

We write the desired solution

$$\bar{u}_b = (\varphi_b^{(s)},\ \varphi_b^{(\eta)},\ \varphi_b^{(\zeta)}) \tag{20}$$

of equation (1) in the form

$$\left.\begin{aligned} \varphi_b^{(l)} &= \exp\left[-i\omega\left(t - \int_0^s \frac{ds}{b(s,\,0,\,0)}\right)\right] V^{(l)}(s,\ \eta,\ \zeta,\ \omega), \\ l &= s,\ \eta,\ \zeta, \end{aligned}\right\} \tag{21}$$

where ω is the frequency and $V^{(l)}$ is a new unknown function which acts as the attenuation function introduced by V.A. Fok. We will assume that ω is fairly large, and we represent $V^{(l)}$ as follows:

$$V^{(l)} = \sum_{k=0}^{k} V^{(l)}\omega^{-k/2}, \quad l = s,\ \eta,\ \zeta. \tag{22}$$

In deriving the parabolic equation we assume that

$$|\eta| = 0(\omega^{-1/2}), \quad |\zeta| = 0(\omega^{-1/2}), \tag{23}$$

$$\left.\begin{aligned} V^{(l)} &= 0(1), \\ \frac{\partial^{e_1+e_2+e_3} V^{(l)}}{\partial s^{e_1}\partial\eta^{e_2}\partial\zeta^{e_3}} &= 0\left(\omega^{\frac{e_2+e_3}{2}}\right), \end{aligned}\right\} \tag{24}$$

where e_1, e_2, e_3 are natural numbers.

We expand the coefficients of (1) in series in powers of η and ζ in the neighborhood of the point $(s, 0, 0)$, allowing for (13) and (15), and then we substitute (21) and (22) for the desired functions, taking into account assumptions (23) and (24).

We will make it our aim to satisfy (1) to within the order ω^1 inclusive. To do this the coefficients of the resultant expansion for identical powers of the parameter $\omega^{1/2}$ are set equal to zero.

We introduce the new variables x_1 and x_2 by the formulas

$$x_1 = \eta\omega^{1/2}, \quad x_2 = \zeta\omega^{1/2}, \tag{25}$$

where, as follows from (23), these new variables are zero-order quantities.

Equation (1) for $r = 1$ yields

$$\left.\begin{aligned} \overset{0}{V}{}^{(s)} &= 0, \\ \overset{1}{V}{}^{(s)} &= ib\left(\frac{\partial \overset{0}{V}{}^{(\eta)}}{\partial x_1} + \frac{\partial \overset{0}{V}{}^{(\zeta)}}{\partial x_2}\right), \end{aligned}\right\} \tag{26}$$

$$\begin{aligned} \overset{2}{V}{}^{(s)} = ib\Bigg\{&\left(\frac{\partial \overset{1}{V}{}^{(\eta)}}{\partial x_1} + \frac{\partial \overset{1}{V}{}^{(\zeta)}}{\partial x_2}\right) \\ &+\left[\frac{\partial}{\partial \eta}\ln\sqrt{g} + \frac{b^2}{a^2 - b^2}\frac{\partial}{\partial\eta}\ln(\mu G_{ss})\right]\overset{0}{V}{}^{(\eta)} \\ &+\left[\frac{\partial}{\partial \zeta}\ln\sqrt{g} + \frac{b^2}{a^2 - b^2}\frac{\partial}{\partial\zeta}\ln(\mu G_{ss})\right]\overset{0}{V}{}^{(\zeta)}\Bigg\} \end{aligned} \tag{27}$$

We may similarly find the subsequent functions $\overset{k}{V}{}^{(s)}$, $k \geqslant 3$. Thus any function $\overset{k}{V}{}^{(s)}$ is defined by the values of the functions $\overset{m}{V}{}^{(\eta)}$ and $\overset{m}{V}{}^{(\zeta)}$ and their derivatives with respect to η and ζ for $m \leqslant k - 1$.

We will now determine the unknown functions $V^{(\eta)}$ and $V^{(\zeta)}$

We introduce the differential operator L:

$$L = \frac{2i}{b}\frac{\partial}{\partial s} + \Delta - \frac{1}{b^2}(K\bar{X}, \bar{X}) + \frac{i}{b}\frac{\partial}{\partial s}\ln(\sqrt{g}\,\mu G^{ss}), \tag{28}$$

where Δ is the Laplace operator

$$\Delta = \frac{\partial^2}{\partial x_1^2} + \frac{\partial^2}{\partial x_2^2},$$

and $\bar{X}$ denotes the vector

$$\bar{X} = x_1\bar{e}_\eta + x_2\bar{e}_\zeta. \tag{29}$$

The matrix K is defined by formula (18).

From equation (1) ($r = 2, 3$; order ω) we obtain

$$L\overset{0}{V}{}^{(\eta)} = 0, \quad L\overset{0}{V}{}^{(\zeta)} = 0. \tag{30}$$

Since the solutions under consideration are concentrated close to L_b, we require that

$$\left.\begin{aligned}&\overset{k}{V}{}^{(l)} \not\equiv 0, \quad \max_{x_1^2+x_2^2=h^2} \left|\overset{k}{V}{}^{(l)}(s,\ x_1,\ x_2)\right| \xrightarrow[h\to\infty]{} 0,\\ &l = s,\ \eta,\ \zeta;\quad k=0,\ 1,\ 2,\ \ldots .\end{aligned}\right\} \tag{31}$$

The solution of the parabolic equations (30) will be sought in the following form:

$$\overset{0}{V}{}^{(l)} = A_0^{(l)}(s)\cdot e^{\frac{i(\Gamma\bar{X},\ \bar{X})}{2b}},\quad l=\eta,\ \zeta. \tag{32}$$

Here $A_0^{(l)}(s)$ are unknown functions, and Γ is a symmetrical matrix:

$$\Gamma = \|\Gamma_{ij}\|,\quad i,\ j=1,\ 2, \tag{33}$$

whose elements Γ_{ij} are functions of the variable s, i.e. $\Gamma_{ij} = \Gamma_{ij}(s)$.

Substituting (32) into (30), we obtain the general Riccati equation for (33):

$$\frac{d\Gamma}{ds} + \Gamma^2 + K - \frac{1}{b}\frac{\partial b}{\partial s}\Gamma = 0, \tag{34}$$

and the following equations for the functions $A_0^{(l)} = A_0^{(l)}(s)$:

$$2\frac{dA_0^{(l)}}{ds} + A_0^{(l)}\frac{\partial}{\partial s}\ln\left(\sqrt{g}\,\mu G^{ss}\right) + A_0^{(l)}\operatorname{tr}\Gamma = 0,\quad l=\eta,\ \zeta, \tag{35}$$

where $\operatorname{tr}\Gamma = \Gamma_{11} + \Gamma_{22}$ is the trace of Γ.

We will seek the solution of (34) in the form

$$\Gamma = \frac{d\gamma}{ds}\cdot\gamma^{-1}, \tag{36}$$

where γ is a new unknown matrix:

$$\gamma = \begin{Vmatrix}\gamma_{11} & \gamma_{12}\\ \gamma_{21} & \gamma_{22}\end{Vmatrix},\quad \gamma_{ij} = \gamma_{ij}(s). \tag{37}$$

Then from (34) we obtain

$$\frac{d^2\gamma}{ds^2} - \frac{1}{b}\frac{\partial b}{\partial s}\cdot\frac{d\gamma}{ds} + K\cdot\gamma = 0. \tag{38}$$

We introduce two vectors $\bar{\gamma}_1$ and $\bar{\gamma}_2$:

$$\left.\begin{aligned}\bar{\gamma}_1 &= \gamma_{11}\bar{e}_\eta + \gamma_{21}\bar{e}_\zeta,\\ \bar{\gamma}_2 &= \gamma_{12}\bar{e}_\eta + \gamma_{22}\bar{e}_\zeta,\end{aligned}\right\} \tag{39}$$

whose components are columns of γ. In accordance with (38), we have

$$\frac{D^2\bar{\gamma}_i}{ds^2}-\frac{1}{b}\frac{\partial b}{\partial s}\cdot\frac{D\bar{\gamma}_i}{ds}+K\bar{\gamma}_i=0, \quad i=1,\ 2. \tag{40}$$

The last equations coincide with (16).

As in [1], we will assume that (16) has four linearly independent solutions, which may be selected in such a way that they form two pairs of complex conjugate vectors:

$$\bar{\gamma}_1,\ \bar{\gamma}_1^*;\quad \bar{\gamma}_2,\ \bar{\gamma}_2^*, \tag{41}$$

which satisfy the following normalization conditions:

$$\left.\begin{aligned}\left(\frac{D\bar{\gamma}_k}{ds},\ \bar{\gamma}_j\right)-\left(\bar{\gamma}_k,\ \frac{D\bar{\gamma}_j}{ds}\right)&=ib\delta_{kj},\\ \left(\frac{D\bar{\gamma}_k}{ds},\ \bar{\gamma}_j^*\right)-\left(\bar{\gamma}_k,\ \frac{D\bar{\gamma}_j^*}{ds}\right)&=0,\quad j,\ k=1,\ 2.\end{aligned}\right\} \tag{42}$$

We note that $\bar{\gamma}_j^*$ is a vector whose components are the complex conjugates of the components of $\bar{\gamma}_j$, $j=1,2$.[1]

Henceforth we will assume that vectors (41) satisfying (42) are known [1].

We take the vectors (41) as the linearly independent solutions of (40) for $i=1,2$. It may be shown that the matrix Γ which is defined by formulas (36), (37) and (39) will be symmetrical, while the matrix

$$\operatorname{Im}\Gamma=\frac{1}{2i}(\Gamma-\Gamma^*) \tag{43}$$

is positive definite [1]. Here and below the matrix A^* is the Hermitian conjugate of A.

Let us return to (35). Substituting the resultant expression for Γ, we find (see the similar considerations in [1])

$$A_0^{(l)}=A_0^{(l)}(s)=\sqrt{\frac{G_{ss}}{\sqrt{g}\,\mu}}\cdot\frac{C^{(l)}}{\sqrt{\det\gamma}},\quad l=\eta,\ \zeta, \tag{44}$$

where $C^{(\eta)}$ and $C^{(\zeta)}$ are arbitrary constants.

Now for the functions $\overset{0}{V}{}^{(l)}$, $l=\eta,\zeta$, we may write

$$\overset{0}{V}{}^{(l)}=C^{(l)}\sqrt{\frac{G_{ss}}{\sqrt{g}\,\mu}}\cdot\frac{1}{\sqrt{\det\gamma}}\cdot\exp\left[\frac{i}{2b}\left(\frac{d\gamma}{ds}\gamma^{-1}X,\ X\right)\right]. \tag{45}$$

---

[1] The scalar product of the vectors $\mathbf{y}=\sum_{i=1}^2 y_i\mathbf{e}_i$ and $\mathbf{z}=\sum_{i=1}^2 z_i\mathbf{e}_i$ has the form $(\mathbf{y},\mathbf{z})=\sum_{i=1}^2 y_iz_i^*$ here.

Formulas (45) specify the principal approximation to the desired solution. Since the matrix Im Γ is positive definite, conditions (31) will hold for the functions $\overset{0}{V}{}^{(\ell)}$, $\ell = s, \eta, \zeta$, (decrease with increasing distance from L_b).

Now to obtain the desired solution which satisfies (1) with the required accuracy (to order ω inclusive), it suffices to limit ourselves in (21) and (22) to terms containing the functions $\overset{0}{V}{}^{(s)}$, $\overset{0}{V}{}^{(\eta)}$, $\overset{0}{V}{}^{(\zeta)}$, $\overset{1}{V}{}^{(s)}$, $\overset{1}{V}{}^{(\eta)}$, $\overset{1}{V}{}^{(\zeta)}$, $\overset{2}{V}{}^{(s)}$ as multipliers. These functions are defined by formulas (26), (27), and (45); any two of the last three of them may be specified arbitrarily, since under the above accuracy requirements they must be related by the single expression (27).

§3. Let us now consider the problem of obtaining the subsequent approximations in constructing the solution (21), (22) of (1) under more stringent accuracy requirements.

As we noted above, the coefficients $\overset{k}{V}{}^{(s)}$ are defined in terms of the functions $\overset{m}{V}{}^{(\eta)}$ and $\overset{m}{V}{}^{(\zeta)}$ for $m < k$; therefore we will confine ourselves here to the determination of these two functions.

We introduce new unknown functions $\overset{k}{W}{}^{(\ell)}$ by the formulas

$$\overset{k}{V}{}^{(l)} = \sqrt{\frac{G_{ss}}{\sqrt{g}\cdot\mu}}\,\overset{k}{W}{}^{(l)}, \quad k = 0,\ 1,\ 2,\ \ldots, \tag{46}$$

and also the operator

$$M = \frac{2i}{b}\frac{\partial}{\partial s} + \Delta - \frac{1}{b^2}(K\bar{X},\ \bar{X}) \tag{47}$$

and the function

$$W = \frac{1}{\sqrt{\det\gamma}}\cdot\exp\left[\frac{i}{2b}(\Gamma\bar{X},\ \bar{X})\right], \tag{48}$$

where, obviously,

$$MW = 0. \tag{49}$$

Below we will use $P_k^{(\ell)}$ and $Q_k^{(\ell)}$, $\ell = \eta, \zeta$, to denote polynomials of degree k of the variables x_1 and x_2 with coefficients which depend on s. These polynomials have a special form. In the first place, they consist only of monomials of odd powers if k is odd, and only of monomials of even powers if k is even. Second, these polynomials are linear homogeneous functions with respect to $\overset{1}{C}{}^{(\eta)}$ and $\overset{1}{C}{}^{(\zeta)}$.

After a number of operations, for the functions $\overset{1}{W}{}^{(\eta)}$ and $\overset{1}{W}{}^{(\zeta)}$ we obtain from (1) ($r = 2, 3$; order $\omega^{1/2}$)

$$M\overset{1}{W}{}^{(l)}=P_3^{(l)}W, \quad l=\eta,\ \zeta, \tag{50}$$

where $P_3^{(l)}$ are certain polynomials which are known at this stage.

It will be shown below that equations (50) have solutions of the form

$$\overset{1}{W}{}^{(l)}=Q_3^{(l)}W, \quad l=\eta,\ \zeta. \tag{51}$$

Allowing for (51), equations (1) ($r = 2, 3$; zero order) yield

$$M\overset{2}{W}{}^{(l)}=P_6^{(l)}W, \quad l=\eta,\ \zeta; \tag{52}$$

the solution of these equations has the form

$$\overset{2}{W}{}^{(l)}=Q_6^{(l)}W, \quad l=\eta,\ \zeta. \tag{53}$$

This process may be continued. Thus in the kth approximation the functions $\overset{k}{W}{}^{(l)}$ will be

$$\overset{k}{W}{}^{(l)}=Q_{3k}^{(l)}W, \quad l=\eta,\ \zeta. \tag{54}$$

In concluding this section, we should note the solution of the equations obtained for the functions $\overset{k}{W}{}^{(l)}$:

$$M\overset{k}{W}=P_{3k}W, \quad k=1,\ 2,\ \ldots, \tag{55}$$

where we have omitted the superscript (l) for the functions $\overset{k}{W}{}^{(l)}$ and the polynomials $P_{3k}^{(l)}$, since the corresponding equations have the same form for the functions $\overset{k}{W}{}^{(\eta)}$ and $\overset{k}{W}{}^{(\zeta)}$. We will show that the solution of the last equation has the form (54).

Following [1], we introduce the auxiliary operators

$$\left.\begin{aligned} \Lambda_j&=\frac{1}{i}(\bar{\gamma}_j,\ \nabla)-\frac{1}{b}\left(\frac{D\bar{\gamma}_j}{dx},\ \bar{X}\right),\\ \Lambda_j^*&=\frac{1}{i}(\bar{\gamma}_j^*,\ \nabla)-\frac{1}{b}\left(\frac{D\bar{\gamma}_j^*}{ds},\ \bar{X}\right),\quad j=1,\ 2, \end{aligned}\right\} \tag{56}$$

where

$$\nabla=\frac{\partial}{\partial x_1}\bar{e}_\eta+\frac{\partial}{\partial x_2}\bar{e}_\zeta.$$

The operators Λ_j, Λ_j^* and M have the following properties:

$$\left.\begin{array}{c}\Lambda_n\Lambda_j-\Lambda_j\Lambda_n=0,\quad \Lambda_n^*\Lambda_j^*-\Lambda_j^*\Lambda_n^*=0,\\ \Lambda_n\Lambda_j^*-\Lambda_j^*\Lambda_n=\delta_{nj},\\ \Lambda_jM-M\Lambda_j=0,\quad \Lambda_j^*M-M\Lambda_j^*=0,\\ \Lambda_jW=0,\quad \Lambda_j^*W\neq 0,\ j,\ n=1,\ 2.\end{array}\right\} \tag{57}$$

Consider the functions

$$u_{\alpha_1\alpha_2}=(\Lambda_1^*)^{\alpha_1}(\Lambda_2^*)^{\alpha_2}W, \tag{58}$$

where α_1 and α_2 are any natural numbers. Applying the operators Λ_2^* and Λ_1^* successively to the function W, we find that

$$u_{\alpha_1\alpha_2}=R_{\alpha_1+\alpha_2}^{(\alpha_1,\,\alpha_1)}\cdot W, \tag{59}$$

where $R_{\alpha_1+\alpha_2}^{(\alpha_1,\alpha_2)}$ is a polynomial of degree $\alpha_1+\alpha_2$ in the variables x_1 and x_2 with coefficients which depend upon s. The polynomials $R_{\alpha_1+\alpha_2}^{(\alpha_1,\alpha_2)}$ will consist only of monomials of even powers if $\alpha_1+\alpha_2$ is even, and of monomials of only odd powers if $\alpha_1+\alpha_2$ is odd. We note that for any α_1 and α_2 we have

$$Mu_{\alpha_1\alpha_2}=0. \tag{60}$$

Integrating by parts and using (57), we may derive the following orthogonality conditions for the functions $u_{\alpha_1\alpha_2}$:

$$(u_{\alpha_1\alpha_2},\ u_{\beta_1\beta_2})=\int_{-\infty}^{+\infty}\int_{-\infty}^{+\infty}u_{\alpha_1\alpha_2}\overline{u_{\beta_1\beta_2}}dx_1dx_2$$

$$=\left|\begin{array}{l}0\quad\text{for}\quad \alpha_1\neq\beta_1\quad\text{or}\quad \alpha_2\neq\beta_2,\\ 2\pi\alpha_1!\,\alpha_2!\quad\text{for}\quad \alpha_1=\beta_1,\ \alpha_2=\beta_2.\end{array}\right\} \tag{61}$$

It follows from the orthogonality relationships that the polynomials $R_{\alpha_1+\alpha_2}^{(\alpha_1,\alpha_2)}$ form a system of linearly independent polynomials. For any fixed degree $\alpha_1+\alpha_2$ there are $\alpha_1+\alpha_2+1$ linearly independent polynomials $R_{\alpha_1+\alpha_2}^{(\alpha_1,\alpha_2)}$. Therefore an arbitrary polynomial of degree k may be expanded in polynomials $R_{\alpha_1+\alpha_2}^{(\alpha_1,\alpha_2)}$, $\alpha_1+\alpha_2\leqslant k$.

Let us reconsider equation (55). We expand the polynomial P_{3k} in the polynomials $R_{\alpha_1+\alpha_2}^{(\alpha_1,\alpha_2)}$:

$$P_{3k}=\sum_{\alpha_1+\alpha_2\leqslant 3k}{}' p_{\alpha_1\alpha_2}R_{\alpha_1+\alpha_2}^{(\alpha_1,\,\alpha_2)}, \tag{62}$$

where $p_{\alpha_1\alpha_2}$ are coefficients of this expansion, which are functions of the variable s.

Multiplying (62) by $W\cdot\overline{u_{\alpha_1\alpha_2}}$ and integrating with respect to x_1 and x_2 over the ranges $(-\infty,+\infty)$, we may write the following analytic expressions for the coefficients $p_{\alpha_1\alpha_2}$:

$$p_{\alpha_1\alpha_2}=\frac{1}{2\pi\alpha_1!\,\alpha_2!}\int_{-\infty}^{+\infty}\int_{-\infty}^{+\infty}P_{3k}\cdot W\cdot\overline{u_{\alpha_1\alpha_2}}dx_1dx_2. \tag{63}$$

We will seek a solution of (55) in the form

$$\overset{k}{W} = \sum_{\beta_1+\beta_2 \leqslant 3k} q_{\beta_1\beta_2}(s)\, u_{\beta_1\beta_2}, \tag{64}$$

where the summation extends only over the pairs β_1, β_2 for which $\beta_1 + \beta_2$ has the same parity as k. In this case the desired formulas (54) will hold for the functions $\overset{k}{W}{}^{(\eta)}$ and $\overset{k}{W}{}^{(\zeta)}$.

Applying the operator M to (64) and allowing for (60), we have

$$\sum_{\alpha_1+\alpha_2 \leqslant 3k} \left(\frac{2i}{b} \frac{dq_{\alpha_1\alpha_2}(s)}{ds} - p_{\alpha_1\alpha_2} \right) u_{\alpha_1\alpha_2} = 0, \tag{65}$$

whence we finally obtain the function

$$q_{\alpha_1\alpha_2} = q_{\alpha_1\alpha_2}(s) = \frac{1}{2i} \int_0^s b p_{\alpha_1\alpha_2} ds. \tag{66}$$

REMARK. It follows from (64) and (65) that a somewhat more general solution of (1) may be obtained. To do this, instead of (66) we take

$$q_{\alpha_1\alpha_2}(s) = \frac{1}{2i} \int_0^s b p_{\alpha_1\alpha_2} ds + q_{\alpha_1\alpha_2}(0)$$

for all α_1 and α_2 with $\alpha_1 + \alpha_2 \leqslant 3k$. Allowing for the arbitrary constants $q_{\alpha_1\alpha_2}(0)$ leads to the fact that the polynomials $P_m^{(l)}$ and $Q_m^{(l)}$ will contain monomials of both even and odd powers. Here the dependence of these polynomials on the quantities $C^{(\eta)}$ and $C^{(\zeta)}$ will be linear.

§4. Consider the rays L_a. The construction of the solutions of (1) which are concentrated along L_a differs from the above method only in some details. We will point out the principal differences.

We replace the transverse velocity b by the longitudinal velocity a in (15), (16) and (18). The desired solution

$$\bar{u}_a = (\varphi_a^{(s)},\ \varphi_a^{(\eta)},\ \varphi_a^{(\zeta)}) \tag{67}$$

may be written in the form

$$\left.\begin{aligned} \varphi_a^{(l)} = \exp\left[-i\omega \cdot \left(t - \int_0^s \frac{ds}{a(s,\ 0,\ 0)} \right] \cdot \left[\sum_{k=0} \overset{k}{V}{}^{(l)}(s,\ \eta,\ \zeta) \cdot \omega^{-k/2} \right], \\ l = s,\ \eta,\ \zeta. \end{aligned}\right\} \tag{68}$$

Assume that the accuracy requirements coincide with the corresponding ones of §2. From (1) ($r = 2, 3$) we obtain

$$\left.\begin{aligned} &\overset{0}{V}{}^{(\eta)} = 0, \quad \overset{1}{V}{}^{(\eta)} = -ia\frac{\partial \overset{0}{V}{}^{(s)}}{\partial x_1}, \quad \overset{2}{V}{}^{(\eta)} = -ia\frac{\partial \overset{1}{V}{}^{(s)}}{\partial x_1} - \\ &-i\frac{a}{a^2-b^2}\left[(a^2-2b^2)\frac{\partial}{\partial \eta}\ln\lambda - b^2\frac{\partial}{\partial \eta}\ln G_{ss}\right]\overset{0}{V}{}^{(s)}, \end{aligned}\right\} \tag{69}$$

$$\left.\begin{aligned} &\overset{0}{V}{}^{(\zeta)} = 0, \quad \overset{1}{V}{}^{(\zeta)} = -ia\frac{\partial \overset{0}{V}{}^{(s)}}{\partial x_2}, \quad \overset{2}{V}{}^{(\zeta)} = -ia\frac{\partial \overset{1}{V}{}^{(s)}}{\partial x_2} - \\ &-i\frac{a}{a^2-b^2}\left[(a^2-2b^2)\frac{\partial}{\partial \zeta}\ln\lambda - b^2\frac{\partial}{\partial \zeta}\ln G_{ss}\right]\overset{0}{V}{}^{(s)}. \end{aligned}\right\} \tag{70}$$

Thus the functions $\overset{m}{V}{}^{(\eta)}$ and $\overset{m}{V}{}^{(\zeta)}$ are defined in terms of $\overset{n}{V}{}^{(s)}$, $n < m$.

Instead of system (30), we arrive at one equation for $\overset{0}{V}{}^{(s)}$:

$$L_a \overset{0}{V}{}^{(s)} = 0, \tag{71}$$

where the operator L_a is obtained from (28) if, as before, the velocity b is replaced by a in this expression. The decrease with increasing distance from L_a will have the form (31).

The process of finding the function $\overset{0}{V}{}^{(s)}$ from (71) is completely analogous to the calculation of $\overset{0}{V}{}^{(\eta)}$ and $\overset{0}{V}{}^{(\zeta)}$ in §2.

To satisfy (1) to within quantities of order ω inclusive, it suffices to confine ourselves in (68) to terms containing the functions $\overset{0}{V}{}^{(l)}$, $\overset{1}{V}{}^{(l)}$ ($l = s, \eta, \zeta$), $\overset{2}{V}{}^{(\eta)}$, $\overset{2}{V}{}^{(\zeta)}$. The last two functions are computed from (69) and (70) in terms of the function $\overset{1}{V}{}^{(s)}$, which may be arbitrarily specified at this stage.

The process of obtaining the subsequent approximations in this case is similar to that of §3. Here, instead of two equations of the form (55), we will have only one equation for the function $\overset{k}{W}{}^{(s)}$, corresponding to the coefficient $\overset{k}{V}{}^{(s)}$. The functions $\overset{k}{V}{}^{(l)}$, $l = \eta, \zeta$, are found in terms of $\overset{m}{V}{}^{(s)}$, $m < k$, from formulas of the form (69) and (70).

§5. We will offer certain conclusions. It follows from (68)–(70) that the components $\varphi_a^{(\eta)}$ and $\varphi_a^{(\zeta)}$ of the vector $\bar{u}_a$ are of order $\omega^{1/2}$, while the component $\varphi_a^{(s)}$ is of zero order. Basically, therefore, in an elastic medium for solutions concentrated near rays L_a, the particles are displaced parallel to the tangent of L_a, and thus longitudinal waves are obtained.

Formulas (21), (22), (26), (27) and (45) show that the components $\varphi_b^{(\eta)}$ and $\varphi_b^{(\zeta)}$ of the vector $\bar{u}_b$, on the other hand, are zero-order quantities, while the component $\varphi_b^{(s)}$ is of order $\omega^{-1/2}$. Basically, therefore, in solutions concentrated close to rays L_b, any particle of the medium is displaced in the plane normal to L_b, and in this case we have transverse waves.

The author wishes to express his sincere thanks to Professor V.M. Babič for formulating the problem and for his guidance.

BIBLIOGRAPHY

1. V.M. Babič, *Eigenfunctions which are concentrated in the neighborhood of a closed geodesic*, Zap. Naučn. Sem. Leningrad. Otdel. Mat. Inst. Steklov. (LOMI) **9** (1968), 15–63. (Russian) MR **39** #3755.

2. M.M. Popov, *Natural oscillations of multi-mirror resonators*, Vestnik Leningrad. Univ. **1969**, no. 22(Fiz. Him. vyp. 4), 42–54. (Russian)

Translated by:
M. Lopez-Morillas

PROPAGATION OF SURFACE WAVES CONCENTRATED NEAR RAYS IN AN INHOMOGENEOUS ELASTIC BODY OF ARBITRARY SHAPE

N. JA. KIRPIČNIKOVA

UDC 517 944/947

Abstract. The paper deals with the construction of asymptotic solutions for the equations of elasticity theory, which are concentrated in the neighborhood of a surface ray L and which propagate along L at a velocity close to the velocity $b(x,y,z)$ of propagation of transverse waves. The surface S on which the ray L lies is taken to be arbitrary but fairly smooth. The resultant oscillations are similar to classical Rayleigh waves, since they occur primarily in the plane which is tangent to L and to the normal to S.

Bibliography: 9 items.

Consider an inhomogeneous elastic body Ω of arbitrary shape, bounded by a fairly smooth stress-free surface S.

Let L be a ray which lies entirely on S. We will understand rays to be extrema of the integral

$$\int \frac{ds}{b(x, y, z)}, \quad (x, y, z) \in S, \tag{1}$$

where ds is an element of arc length of the curve along which the integration is taken and $b(x,y,z)$ is the transverse velocity of propagation of elastic waves in Ω.

The aim of this paper is to determine the high-frequency asymptotic behavior of the solutions of the equations of elasticity theory which are concentrated in the neighborhood of L; the nature of these solutions is that of waves which propagate along L at a velocity close to $b(x,y,z)$. As opposed to Love waves, which also possess these properties, we introduce the requirement that the displacement of particles within Ω in the zero approximation be parallel to the plane of the normal section of S along L.

The displacement vector $\mathbf{u}$ for the elastic medium is determined as the sum of the longitudinal and transverse waves $\mathbf{u}_a$ and $\mathbf{u}_b$, which satisfy the conditions for there to be no stress at the boundary. The results indicate that the longitudinal wave falls off immediately as we go from the boundary into the body. The transverse wave, on the other hand, is concentrated in a solid waveguide close to the surface along L. Its behavior in the layer close to the surface corresponds to eigenfunctions of the "whispering gallery" type.

AMS 1970 *subject classifications*. Primary 73D99; 35B40.

Within Ω, the transverse waves $\mathbf{u}_b$ are asymptotically decisive. These waves are similar to Love waves, but the particles are displaced primarily in the direction of the normal to the boundary of S, and not perpendicular to the plane of normal section of the surface along L, as for Love waves.

The order of the longitudinal and transverse waves on S turns out to be the same. For both $\mathbf{u}_a$ and $\mathbf{u}_b$, the particles are displaced parallel to the plane of the normal section of the surface corresponding to L. This property permits us to compare these waves on S with Rayleigh waves.

The resultant asymptotic behavior is a generalization of the corresponding results of [1] to the case of an elastic body of arbitrary shape which is inhomogeneous with respect to its three coordinates. Unlike [1], however, the longitudinal and transverse components of $\mathbf{u}$, as functions of the variable α, are similar to eigenfunctions of the "bouncing ball" type [2–4], i.e. they are concentrated along L in a narrow band in terms of the coordinate α and decrease exponentially outside this band.

§1. Let q^1, q^2, q^3 be curvilinear coordinates in Ω. In this coordinate system the equations of elasticity theory have the form

$$\left.\begin{aligned} \sigma_{ij}\sqrt{g}\,\frac{\partial G^{ij}}{\partial q^s} + 2\frac{\partial}{\partial q^i}(\sigma_{sj}G^{ij}\sqrt{g}) - 2\rho\sqrt{g}\,G_{is}\frac{\partial^2\varphi^{(i)}}{\partial t^2} = 0,\\ i,\ j,\ s = 1,\ 2,\ 3. \end{aligned}\right\} \tag{2}$$

Here and henceforth, repeating indices in each product are summation indices. The σ_{ij} in (2) denote the components of the stress tensor. For an ideal isotropic medium, these components are related to the components of the displacement vector $\mathbf{u} = (\varphi^{(1)}, \varphi^{(2)}, \varphi^{(3)})$ as follows:

$$\begin{aligned} \sigma_{ij} &= \frac{\lambda}{\sqrt{g}}\frac{\partial}{\partial q^s}(\varphi^{(s)}\sqrt{g})\,G_{kj}\delta^k_i \\ &\quad + \mu\left[\frac{\partial G_{ij}}{\partial q^s}\varphi^{(s)} + G_{is}\frac{\partial\varphi^{(s)}}{\partial q^j} + G_{js}\frac{\partial\varphi^{(s)}}{\partial q^i}\right], \quad k = 1,\ 2,\ 3, \end{aligned} \tag{3}$$

where δ^k_i is the Kronecker delta; $\lambda(q^1, q^2, q^3)$ and $\mu(q^1, q^2, q^3)$ are the Lamé parameters; $\rho(q^1, q^2, q^3)$ is the density of the medium. An element of length ds in the (q^1, q^2, q^3) coordinate system is defined by the expression

$$ds^2 = G_{ij}dq^i dq^j,$$

where the G_{ij} are the components of the metric tensor in the space of the coordinates (q^1, q^2, q^3); $g = \det\|G_{ij}\|$; the matrix $\|G_{ij}\|$ is the inverse of $\|G^{ij}\|$; t is time. The functions λ, μ and ρ are assumed to be differentiable a sufficient number of times.

Let L be a ray, specified on S, which is an extremum of integral (1), and let $b = \mu/\rho$.

As the coordinate line (q^1) we take the ray L to be such that the parameter $q^1 = \tau$ characterizes points on the ray, $q^2 = \alpha$ is the distance to L along the extremum of (1) taken orthogonally to a ray on S, while the parameter $q^3 = \nu$ is the distance from an arbitrary point of Ω to S along the normal to S such that internal points of Ω correspond to values $\nu > 0$.

The radius vector $\mathbf{X}$ of an arbitrary point M near the surface S of Ω has the form

$$\mathbf{X} = \mathbf{r}(\tau, \alpha) + \nu \mathbf{n}(\tau, \alpha),$$

where $\mathbf{r} = \mathbf{r}(\tau, \alpha)$ is the parametric equation of S; $\mathbf{n}(\tau, \alpha)$ is the unit vector of the normal to S and is directed toward the interior of the body.

Since $ds^2 = (d\mathbf{X}, d\mathbf{X})$, we obtain the following for the coefficients of the metric tensor:

$$\begin{gathered}
G_{\tau\tau} = (\bar{X}'_\tau, \bar{X}'_\tau) = g_{\tau\tau} - 2\nu b_{\tau\tau} + \nu^2 (\mathbf{n}'_\tau, \mathbf{n}'_\tau), \\
G_{\tau\alpha} = (\bar{X}'_\tau, \bar{X}'_\alpha) = g_{\alpha\tau} - 2\nu b_{\tau\alpha} + \nu^2 (\mathbf{n}'_\tau, \mathbf{n}'_\alpha), \\
G_{\alpha\alpha} = (\bar{X}'_\alpha, \bar{X}'_\alpha) = g_{\alpha\alpha} - 2\nu b_{\alpha\alpha} + \nu^2 (\mathbf{n}'_\alpha, \mathbf{n}'_\alpha), \\
G_{\alpha\nu} = 0, \quad G_{\tau\nu} = 0, \quad G_{\nu\nu} = 1.
\end{gathered}$$

All of our subsequent analysis will be close to L, i.e. at fairly small values of α and ν, and therefore, neglecting terms of order $|\alpha^3|$, $|\alpha|^2\nu$, $|\alpha|\nu^2$ and ν^3, we obtain

$$\left.\begin{aligned}
G_{\tau\tau} &= g_{\tau\tau} + \frac{\partial g_{\tau\tau}}{\partial \alpha}\alpha - 2b_{\tau\tau}\nu + \frac{1}{2}\frac{\partial^2 g_{\tau\tau}}{\partial \alpha^2}\alpha^2 \\
&\qquad - 2\frac{\partial b_{\tau\tau}}{\partial \alpha}\nu\alpha + (\bar{n}'_\tau \cdot \bar{n}'_\tau)\nu^2 + 0(|\alpha|^3, |\alpha|^2\nu, |\alpha|\nu^2, \nu^3), \\
G_{\tau\alpha} &= -2b_{\tau\alpha}\nu - 2\frac{\partial b_{\tau\alpha}}{\partial \alpha}\nu\alpha + (\bar{n}'_\tau \cdot \bar{n}'_\alpha)\nu^2 + 0(|\alpha|^3, |\alpha|^2\nu, |\alpha|\nu^2, \nu^3), \\
G_{\alpha\alpha} &= g_{\alpha\alpha} + \frac{\partial g_{\alpha\alpha}}{\partial \alpha}\cdot\alpha - 2b_{\alpha\alpha}\cdot\nu + \frac{1}{2}\frac{\partial^2 g_{\alpha\alpha}}{\partial \alpha^2}\alpha^2 - 2\frac{\partial b_{\alpha\alpha}}{\partial \alpha}\nu\alpha \\
&\qquad + (\bar{n}'_\alpha \cdot \bar{n}'_\alpha)\nu^2 + 0(|\alpha|^3, |\alpha|^2\nu, |\alpha|\nu, \nu^3), \\
G_{\tau\nu} &= 0, \quad G_{\alpha\nu} = 0, \quad G_{\nu\nu} = 1,
\end{aligned}\right\} \tag{4}$$

where $g_{\tau\tau}$, $g_{\tau\alpha}$, $g_{\alpha\alpha}$ and $b_{\tau\tau}$, $b_{\tau\alpha}$, $b_{\alpha\alpha}$ are respectively the coefficients of the first and second quadratic Gaussian forms for S. We will show that all the coefficients for different powers of α and ν in (4) are calculated at the point $(\tau, 0, 0)$. The (τ, α, ν) coordinate system is orthogonal on S, and therefore $g_{\tau\alpha} = 0$.

Consider integral (1) on S in the (τ, α, ν) system:

$$\int \frac{ds}{b(x, y, z)} = \int \sqrt{\frac{g_{\tau\tau}}{b^2} + \frac{g_{\alpha\alpha}}{b^2}\cdot\alpha'_\tau}\cdot d\tau.$$

The equation $\alpha = \alpha(\tau)$ defines the extrema of this integral. Since $\alpha = 0$ is the equation of L, which is also an extremum, the Euler equation implies for (1) that

$$\frac{\partial}{\partial \alpha}\left(\frac{g_{\tau\tau}}{b^2}\right)\Big|_{\alpha=0} = 0. \tag{5}$$

We will make important use of this condition in the intermediate calculations below.

Since we are constructing solutions which oscillate near L on the surface and attenuate exponentially with increasing distance from S, we must require that the following inequality hold:[1)]

$$k(\tau)+\frac{1}{b}\frac{\partial b}{\partial \nu}\Big|_{\alpha=0,\ \nu=0}>0, \tag{6}$$

which may be written somewhat differently:

$$\frac{\partial}{\partial \nu}\left(\frac{G_{\tau\tau}}{b^2}\right)\Big|_{\alpha=0,\ \nu=0}<0.$$

In (6) we have used $k(\tau)$ to denote the curvature of the normal section of S corresponding to the ray L at the point with coordinates $(\tau, 0, 0)$; we have

$$k(\tau)=\frac{b_{\tau\tau}}{b^2}=-\frac{(\mathbf{r}'_\tau,\ \mathbf{n}'_\tau)}{b^2},$$

where $\mathbf{r}'_\tau/b$ is the unit vector of the tangent to L and is directed towards increasing τ. The quantity

$$\left(k+\frac{1}{b}\frac{\partial b}{\partial \nu}\right)^{-1}$$

acts as the effective radius of curvature of the normal section of S corresponding to the curve L [2]. If the velocity $b(x,y,z)$ is constant, condition (6) means that the center of curvature of the normal section of S along the curve lies on the side where $\nu>0$. Hence the concentration effect of the solution in the layer taken in terms of ν appears inside the body.

The boundary conditions for the absence of stress on the boundary S of Ω can be written as follows:

$$\sigma_{i\nu}\big|_{\nu=0}=0, \quad i=\tau,\ \alpha,\ \nu, \tag{7}$$

where the components σ_{ij} of the stress tensor are given by (3).

§2. Let us now select the general form of the high-frequency asymptotic behavior of the solution of our problem. We note that certain procedures of Babič and Lazutkin [3] will be used here.

Assume that the desired solution

$$\bar{u}=(\varphi^{(\tau)},\ \varphi^{(\alpha)},\ \varphi^{(\nu)}) \tag{8}$$

oscillates at a fairly high frequency ω $(\omega \to \infty)$. We can write the components of the displacement vector in the form

---

1) As will be seen from the formulas obtained below, inequality (6) has a clear physical meaning. When it holds, waves of the "whispering gallery" type concentrate inside Ω near S.

$$\varphi^{(\tau)} = \exp\left[-i\omega(t-\tau) + i\sum_{k=-5}^{0} \alpha_k(\tau, \eta, \zeta)\,\omega^{-\frac{k}{6}}\right]$$
$$\times\{\exp[-\omega M(\tau, \eta)\nu]\,V_1^{(\tau)}(\tau, \eta, \zeta, \omega) + V_2^{(\tau)}(\tau, \eta, \zeta, \omega)\}, \tag{9}$$

$$\varphi^{(\alpha)} = \exp\left[-i\omega(t-\tau) + i\sum_{k=-5}^{0} \alpha_k(\tau, \eta, \zeta)\,\omega^{-k/6}\right]$$
$$\times\{\exp[-\omega M(\tau, \eta)\cdot\nu]\,V_1^{(\alpha)}(\tau, \eta, \zeta, \omega) + V_2^{(\alpha)}(\tau, \eta, \zeta, \omega)\}, \tag{10}$$

$$\varphi^{(\nu)} = \exp\left[-i\omega(t-\tau) + i\sum_{k=-5}^{0} \alpha_k(\tau, \eta, \zeta)\,\omega^{-k/6}\right]$$
$$\times\{\exp[-\omega M(\tau, \eta)\cdot\nu]\,V_1^{(\nu)}(\tau, \eta, \zeta, \omega) + V_2^{(\nu)}(\tau, \eta, \zeta, \omega)\}. \tag{11}$$

Here t is the time, $M(\tau,\eta)$ is an unknown real function, and the variables η and ζ are defined as follows:

$$\eta = \alpha\omega^{1/2}, \qquad \zeta = \nu\omega^{2/3}. \tag{12}$$

Assuming that $\alpha = 0(\omega^{-1/2})$ and $\nu = 0(\omega^{-2/3})$, we obtain that η and ζ are quantities of order 0(1).

We expand the functions $V_\kappa^{(i)}$ $(i = \tau, \alpha, \nu; \kappa = 1, 2)$ in a series in the small parameter $\omega^{-1/6}$:

$$V_\chi^{(i)} = \sum_{k=0} \overset{k}{V}{}_\chi^{(i)}(\tau, \eta, \zeta)\cdot\omega^{-k/6}. \tag{13}$$

In this paper, as in [4], it is possible to choose certain unknown functions in the form of polynomials in the variables η and ζ.

We write the functions α_k, $\overset{k}{V}{}_1^{(i)}$ $(i = \tau, \alpha, \nu)$ as follows:

$$\alpha_k = \sum_{l+m=0}^{P_k} \alpha_{k\,lm}(\tau)\,\eta^l\zeta^m, \tag{14}$$

$$\overset{k}{V}{}_1^{(i)} = \overset{k}{V}{}_{1,1}^{(i)}\cdot D_q(\sqrt{2}\,\eta\chi) + \overset{k}{V}{}_{1,2}^{(i)}\cdot D_q'(\sqrt{2}\,\eta\chi), \tag{15}$$

where $\overset{k}{V}{}_{1,1}^{(i)}$, $\overset{k}{V}{}_{1,2}^{(i)}$ are polynomials in η and ζ with coefficients which depend on τ. The quantity P_k in (14) denotes the degree of the corresponding polynomial.

We will seek the functions $\overset{k}{V}{}_2^{(i)}$ $(i = \tau, \alpha, \nu)$ in a somewhat more general form than in [1, 3]. We will give only the expression for $\overset{k}{V}{}_2^{(\nu)}$:

$$\overset{k}{V}{}_2^{(\nu)} = [A_k(\tau, \eta, \zeta)\cdot v(\psi\zeta - \xi_0) + B_k(\tau, \eta, \zeta)\cdot v'(\psi\zeta - \xi_0)]\,D_q(\sqrt{2}\,\eta\chi)$$
$$+ [C_k(\tau, \eta, \zeta)\cdot v(\psi\zeta - \xi_0) + D_k(\tau, \eta, \zeta)\cdot v'(\psi\zeta - \xi_0)]\,D_q'(\sqrt{2}\,\eta\chi)$$
$$+ \overset{k}{W}{}^{(\nu)}(\tau, \eta, \zeta)\,D_q(\sqrt{2}\,\eta\chi) + \overset{k}{U}{}^{(\nu)}(\tau, \eta, \zeta)\,D_q'(\sqrt{2}\,\eta\chi). \tag{16}$$

The quantities $A_k, \ldots, D_k$ in (16) are polynomials of the form (14), ξ_0 is a positive number to be determined; $\psi = \psi(\tau)$ and $\chi = \chi(\tau)$ are unknown real functions; ω is the frequency. The function $v(x)$ is the solution of the Airy equation which has the following asymptotic behavior [5]:

$$v(x) \sim \frac{1}{2} x^{-1/4} \exp\left(-\frac{2}{3} x^{3/2}\right), \quad x \to \infty.$$

The functions $D_q(\sqrt{2}\, z)$ are familiar parabolic cylinder functions [6] with integer index $q = 0, 1, 2, \ldots$:

$$D_q(\sqrt{2}\, z) = \exp\left(-\frac{z^2}{2}\right) \cdot H_q(z) \cdot 2^{-q/2}, \tag{17}$$

where $H_q(z) = (-1)^q e^{z^2} d^q e^{-z^2}/dz^q$ are Hermitian polynomials. $\overset{k}{W}{}^{(\nu)}$ and $\overset{k}{U}{}^{(\nu)}$ denote unknown functions; it is assumed that they are linearly independent with the Airy functions and their derivatives.

§3. It will be our aim to satisfy equation (2) with accuracy to $0(\omega^{5/6})$. In particular, in each of the unknown expansions we will find the two first coefficients which are not identically zero for the corresponding powers of the parameter $\omega^{-1/6}$.

We expand the coefficients of (2) and of boundary conditions (7) in series in powers of α and ν in the vicinity of the point $(\tau, 0, 0)$, taking (4) and (5) into account. Then we substitute solution (9)–(11) into (2) and (7).

It may be shown that for $\nu > 0$ equation (2) should be solved independently for the terms of (9)–(11) containing the functions V_1 and V_2 respectively as factors. But in the desired solution these components will be dependent, since they are of equivalent importance in boundary conditions (7).

As noted above, the unknown functions of the problem are $\overset{k}{V}{}_1^{(i)}$, $\overset{k}{V}{}_2^{i}$, α_k, $\psi(\tau)$, $\chi(\tau)$, $M(\tau, \eta)$. We will first determine $\overset{k}{V}{}_1^{(i)}$ and $M(\tau, \eta)$.

We equate to zero the coefficients of (2) $(s = 1, 3)$ and of (7) $(i = \tau, \nu)$ for the same powers of $\omega^{1/6}$.

The boundary conditions of orders $0(\omega)$ and $0(\omega^{5/6})$ and the corresponding equations for the functions $\overset{0}{V}{}_1^{(\tau)}$, $\overset{0}{V}{}_1^{(\nu)}$, $\overset{1}{V}{}_1^{(\tau)}$ and $\overset{1}{V}{}_1^{(\nu)}$ yield

$$\overset{0}{V}{}_1^{(\tau)} = 0, \quad \overset{0}{V}{}_1^{(\nu)} = 0, \quad \overset{1}{V}{}_1^{(\tau)} = 0, \quad \overset{1}{V}{}_1^{(\nu)} = 0. \tag{18}$$

The coefficients $\overset{k}{V}{}_1^{(\tau)}$ and $\overset{k}{V}{}_1^{(\nu)}$ $(k = 2, 3, 4, 5)$ are determined from the following recurrence relationship:

$$\left.\begin{aligned} &a_{11}\overset{k}{V}{}_1^{(\tau)} + a_{21}\overset{k}{V}{}_1^{(\nu)} - 2Mb^4 \frac{\partial \overset{k-2}{V}{}_1^{(\tau)}}{\partial \zeta} + i(a^2 - b^2)\frac{\partial \overset{k-2}{V}{}_1^{(\nu)}}{\partial \zeta} = 0, \\ &a_{12}\overset{k}{V}{}_1^{(\tau)} + a_{22}\overset{k}{V}{}_1^{(\nu)} + i(a^2 - b^2)\frac{\partial \overset{k-2}{V}{}_1^{(\tau)}}{\partial \zeta} - 2Ma^2 \frac{\partial \overset{k-2}{V}{}_1^{(\nu)}}{\partial \zeta} = 0, \end{aligned}\right\} \tag{19}$$

where

$$\left.\begin{aligned} a_{11} &= g_{\tau\tau}(1 - b^2M^2) - a^2, \\ a_{12} &= a_{21} = -i(a^2 - b^2)M, \\ a_{22} &= a^2M^2, \end{aligned}\right\} \tag{20}$$

$$a^2 = \frac{\lambda + 2\mu}{\rho}, \tag{21}$$

and a denotes the longitudinal velocity of propagation of elastic waves.

The condition that system (19) be solvable for $k = 2$ and the condition $M > 0$ (the quantity $\exp(-\omega M\nu)$ should decrease with increasing ν) specify the function $M(\tau, \eta)$ on L as follows:

$$M(\tau, \eta)|_L = M(\tau, 0) = \sqrt{\frac{1}{g_{\tau\tau}} - \frac{1}{a^2}}\,\Bigg|_L. \tag{22}$$

In view of the fact that the functions $\overset{k}{V}{}_1^{(\tau)}$ and $\overset{k}{V}{}_1^{(\nu)}$ have been taken to be polynomials in η and ζ with coefficients which depend on τ, it suffices to make them subject to the expressions

$$\left.\begin{aligned} \overset{k}{V}{}_1^{\nu} &= iMb^2\overset{k}{V}{}_1^{(\tau)}, \\ \frac{\partial \overset{k-2}{V}{}_1^{(\tau)}}{\partial\zeta} &= \frac{\partial \overset{k-2}{V}{}_1^{(\nu)}}{\partial\zeta} = 0, \quad k = 2, 3, 4, 5, \end{aligned}\right\} \tag{23}$$

in order to satisfy (19).

The boundary conditions yield

$$\overset{k}{V}{}_1^{(\tau)}|_{\zeta=0} = 2i\frac{\partial \overset{k-2}{V}{}_2^{(\nu)}}{\partial\zeta}\Bigg|_{\zeta=0}, \quad k = 2, 3, 4, 5. \tag{24}$$

The equations for the functions $\overset{k}{V}{}_1^{(\tau)}$ and $\overset{k}{V}{}_1^{(\nu)}$ $(k = 6, 7, \dots)$ may be conveniently written as

$$\left.\begin{aligned} a_{11}\overset{k}{V}{}_1^{(\tau)} + a_{21}\overset{k}{V}{}_1^{(\nu)} &= f_1^{(k)}, \\ a_{21}\overset{k}{V}{}_1^{(\tau)} + a_{22}\overset{k}{V}{}_1^{(\nu)} &= f_2^{(k)}, \quad k = 6, 7, \dots, \end{aligned}\right\} \tag{25}$$

where $f_1^{(k)}$ and $f_2^{(k)}$ are certain functions of $\overset{m}{V}{}_1^{(\tau)}$ and $\overset{m}{V}{}_1^{(\nu)}$ for $m \leqslant k - 4$ and of their derivatives for $m \leqslant k - 2$. The functions $f_1^{(6)}$ and $f_2^{(6)}$ are as follows:

$$\left.\begin{aligned} f_1^{(6)} &= 2a^2\frac{\partial\alpha_{-2}}{\partial\tau}\overset{2}{V}{}_1^{(\tau)} + 2Mb^2g_{\tau\tau}\frac{\partial\overset{4}{V}{}_1^{(\tau)}}{\partial\zeta} + i(a^2 - b^2)\left(M\frac{\partial\alpha_{-2}}{\partial\tau}\overset{2}{V}{}_1^{(\nu)} - \frac{\partial\overset{4}{V}{}_1^{(\nu)}}{\partial\zeta}\right), \\ f_2^{(6)} &= i(a^2 - b^2)\left(M\frac{\partial\alpha_{-2}}{\partial\tau}\overset{2}{V}{}_1^{(\tau)} - \frac{\partial\overset{4}{V}{}_1^{(\tau)}}{\partial\zeta}\right) + 2\frac{\partial\alpha_{-2}}{\partial\tau}\overset{2}{V}{}_1^{(\nu)} + 2a^2M\frac{\partial\overset{4}{V}{}_1^{(\nu)}}{\partial\zeta}. \end{aligned}\right\} \tag{26}$$

The requirement that system (25) be solvable leads to the conditions

$$f_1^{(k)} + iMb^2 f_2^{(k)} = 0, \quad k = 6,\ 7,\ \ldots \tag{27}$$

For $k = 6$ the corresponding condition can be rewritten as

$$\frac{\partial \overset{4}{V}{}^{(\tau)}}{\partial \zeta} = -\frac{1}{Mb^2}\,\frac{\partial \alpha_{-2}}{\partial \tau}\,\overset{2}{V}{}_1^{(\tau)}, \tag{28}$$

and we can specify the dependence between the desired functions $\overset{6}{V}{}_1^{(\tau)}$ and $\overset{6}{V}{}_1^{(\nu)}$ by the expression

$$\overset{6}{V}{}_1^{(\nu)} = iMb^2\left[\overset{6}{V}{}_1^{(\tau)} + \frac{\partial \alpha_{-2}}{\partial \tau}\,\overset{2}{V}{}_1^{(\tau)}\right]. \tag{29}$$

We note that the functions $\partial\alpha_{-2}/\partial\tau$ and $\overset{6}{V}{}_1^{(\tau)}$ in (29) were left undefined. The first of these will be defined below (see §4), while the second can be easily found from the boumdary conditions and the conditions (27) that system (25) be solvable for $k = 8$.

We may similarly compute all the functions $\overset{k}{V}{}_1^{(\nu)}$ and $\overset{k}{V}{}_1^{(\tau)}$ for $k = 7, 8, \ldots$.

In order to satisfy equations (2) ($s = 1, 3$) with accuracy to order $0(\omega^{5/6})$ it suffices to find the coefficients $\overset{k}{V}{}_1^{(\tau)}$ and $\overset{k}{V}{}_1^{(\nu)}$ up to the index $k = 6$ inclusive.

It follows from (2) ($s = 2$) that when the expressions

$$\left.\begin{aligned} &\overset{k}{V}{}_1^{(\alpha)} = 0, \quad k = 0,\ 1,\ 2,\ 3,\ 4,\\ &\overset{m}{V}{}_1^{(\alpha)} = -\frac{a^2}{b^2}\left(i\,\frac{\partial \overset{m-3}{V}{}_1^{(\tau)}}{\partial \eta} - M\,\frac{\partial \overset{m-3}{V}{}_1^{(\nu)}}{\partial \eta}\right), \quad m = 5, 6, \end{aligned}\right\} \tag{30}$$

hold, it will also be satisfied up to order $0(\omega^{5/6})$.

§4. Let us now determine the as yet unknown functions α_k, $\overset{k}{V}_2$, ψ and χ and the number ξ_0.

Equation (2) ($s = 1$) yields expressions for the functions $\overset{k}{V}{}_2^{(\tau)}$. Equating the coefficients for different powers of $\omega^{1/6}$ to zero, beginning with ω^2 to ω^1 inclusive, we obtain

$$\left.\begin{aligned} &\overset{0}{V}{}_2^{\tau} = 0, \quad \overset{1}{V}{}_2^{\tau} = 0, \quad \overset{k}{V}{}_2^{\tau} = i\left(\frac{\partial \overset{k-2}{V}{}_2^{(\nu)}}{\partial \zeta} + i\,\frac{\partial \alpha_0}{\partial \eta}\,\overset{k-3}{V}{}_2^{\alpha} + \frac{\partial \overset{k-3}{V}{}_2^{\alpha}}{\partial \eta}\right), \quad k = 2,\ 3,\ 4,\ 5,\\ &\overset{6}{V}{}_2^{\tau} = i\,\frac{\partial \overset{4}{V}{}_2^{\nu}}{\partial \zeta} + i\,\frac{\partial \overset{3}{V}{}_2^{\alpha}}{\partial \eta} + i\left[\frac{\partial}{\partial \nu}\ln\sqrt{g} + \frac{b^2}{(a^2 - b^2)}\,\frac{\partial}{\partial \nu}\ln(\mu G_{\tau\tau})\right]\overset{0}{V}{}_2^{\nu}\\ &\quad - \frac{\partial \alpha_0}{\partial \eta}\,\overset{3}{V}{}_2^{\alpha} + i\left[\frac{\partial}{\partial \alpha}\ln\sqrt{g} + \frac{b^2}{(a^2 - b^2)}\,\frac{\partial}{\partial \alpha}\ln(\mu G_{\tau\tau})\right]\overset{0}{V}{}_2^{\alpha}\\ &\quad + ib^2\psi^2\xi_0\,\frac{\partial \overset{0}{V}{}_2^{\nu}}{\partial \zeta} + i\,\frac{b^4\psi^3}{a^2 - b^2}\,\overset{0}{V}{}_2^{\nu}. \end{aligned}\right\} \tag{31}$$

We may similarly find the subsequent coefficients $\overset{k}{V}{}_2^{(\tau)}$ for $k > 6$. It is easy to see that any coefficient $\overset{k}{V}{}_2^{(\tau)}$ is completely defined by the values of $\overset{m}{V}{}_2^{(\nu)}$ and $\overset{n}{V}{}_2^{(\alpha)}$ for $m \leqslant k - 2$ and $n \leqslant k - 3$.

Equating the coefficients for identical powers of $\omega^{1/6}$ to zero, beginning with ω^2, in equations (2) ($s = 2, 3$) we will obtain expressions from which the functions α_k, $\overset{k}{V}{}_2^{(\nu)}$ and $\overset{k}{V}{}_2^{(\alpha)}$ may be found. For the α_k we find

$$\left.\begin{aligned} &\frac{\partial \alpha_k}{\partial \zeta} = 0, \quad k = -5, \ -4, \ -3, \ -2, \ -1, \ 0, \\ &\frac{\partial \alpha_l}{\partial \eta} = 0, \quad l = -5, \ -4, \ -3, \ -2, \ -1, \\ &\frac{\partial \alpha_m}{\partial \tau} = 0, \quad m = -5, \ -4, \ -3. \end{aligned}\right\} \tag{32}$$

Equation (2) ($s = 3$; order $\omega^{4/3}$) yields

$$(a^2 - b^2)\, i \frac{\partial \overset{0}{V}{}_2^{(\tau)}}{\partial \zeta} - 2 \frac{\partial \alpha_{-2}}{\partial \tau} \overset{0}{V}{}_2^{(\nu)} - \frac{\partial}{\partial \nu}(b^2 G^{\tau\tau}) \cdot \zeta \overset{0}{V}{}_2^{(\nu)} + a^2 \frac{\partial^2 \overset{0}{V}{}_2^{(\nu)}}{\partial \zeta^2} = 0. \tag{33}$$

Allowing for (16) and (31), we obtain the functions $\psi = \psi(\tau)$ and $\alpha_{-2} = \alpha_{-2}(\tau)$ in the form

$$\psi^3 = -\frac{1}{b^2} \frac{\partial}{\partial \nu} \left(\frac{G_{\tau\tau}}{b^2} \right) \Big|_{\nu = 0, \ \alpha = 0}, \tag{34}$$

$$\alpha_{-2} = -\frac{\xi_0}{2} \int_0^\tau \psi^2(\tau)\, b^2(\tau)\, d\tau. \tag{35}$$

When applied to formula (34), inequality (6) may be rewritten in the form

$$\psi(\tau) > 0. \tag{36}$$

The function $\overset{0}{V}{}_2^{\nu}$ will satisfy the following equation and boundary condition:

$$\left.\begin{aligned} &\frac{\partial^2 \overset{0}{V}{}_0^{\nu}}{\partial \zeta^2} - \psi^2 (\psi \zeta - \xi_0) \overset{0}{V}{}_2^{\nu} = 0, \\ &\overset{0}{V}{}_2^{\nu}\big|_{\zeta = 0} = 0, \quad \overset{0}{V}{}_2^{\nu} \xrightarrow[\zeta \to \infty]{} 0. \end{aligned}\right| \tag{37}$$

For $\overset{0}{V}{}_2^{\alpha}$ we similarly obtain

$$\left.\begin{aligned} &\frac{\partial^2 \overset{0}{V}{}_2^{\alpha}}{\partial \zeta^2} - \psi^2 (\psi \zeta - \xi_0) \overset{0}{V}{}_2^{\alpha} = 0, \\ &\frac{\partial \overset{0}{V}{}_2^{\alpha}}{\partial \zeta}\Big|_{\zeta = 0} = 0, \quad \overset{0}{V}{}_2^{\alpha} \xrightarrow[\zeta \to \infty]{} 0. \end{aligned}\right| \tag{38}$$

Since for the desired solution the particles are basically displaced parallel to the plane of normal section of the surface along L, it is necessary that

$$\overset{0}{V}{}^{\alpha}_{2} \equiv 0, \quad \overset{0}{V}{}^{\nu}_{2} \neq 0. \tag{39}$$

Hence the function $\overset{0}{V}{}^{\nu}_{2}$ is defined as follows:

$$\left.\begin{aligned} \overset{0}{V}{}^{\nu}_{2} &= H_0(\tau, \eta)\, v(\psi\zeta - \xi_0), \\ H_0 &= A_0 D_q + C_0 D'_q, \end{aligned}\right\} \tag{40}$$

$$A_0 = A_0(\tau, \eta), \quad C_0 = C_0(\tau, \eta), \quad B_0 \equiv 0, \quad D_0 \equiv 0, \quad \overset{0}{W}{}^{\nu} = 0, \quad \overset{0}{U}{}^{\nu} = 0. \tag{41}$$

From the limiting conditions of problem (37) we find that

$$v(-\xi_0) = 0, \quad \xi_0 = x_p, \quad p = 1, 2, 3, \ldots, \tag{42}$$

where x_p are the roots of the Airy functions. We note that all the x_p are positive numbers.

The functions $\overset{1}{V}{}^{\nu}_{2}$ and $\overset{1}{V}{}^{\alpha}_{2}$, like $\overset{0}{V}{}^{\nu}_{2}$ and $\overset{0}{V}{}^{\alpha}_{2}$, satisfy the Sturm-Liouville problem:

$$\left.\begin{aligned} &\frac{\partial^2 \overset{1}{V}{}^{\nu}_{2}}{\partial \zeta_1^2} - (\zeta_1 - \xi_0)\, \overset{1}{V}{}^{\nu}_{2} = -\frac{1}{b^2\psi^2} \cdot 2\, \frac{\partial \alpha_{-1}}{\partial \tau}\, \overset{0}{V}{}^{\nu}_{2}, \\ &\overset{1}{V}{}^{\nu}_{2}\Big|_{\zeta_1 = 0} = 0, \quad \overset{1}{V}{}^{\nu}_{2} \xrightarrow[\zeta_1 \to \infty]{} 0; \end{aligned}\right| \tag{43}$$

$$\left.\begin{aligned} &\frac{\partial^2 \overset{1}{V}{}^{\alpha}_{2}}{\partial \zeta_1^2} - (\zeta_1 - \xi_0)\, \overset{1}{V}{}^{\alpha}_{2} = 0, \\ &\frac{\partial \overset{1}{V}{}^{\alpha}_{2}}{\partial \zeta_1}\Bigg|_{\zeta_1 = 0} = 0, \quad \overset{1}{V}{}^{\alpha}_{2} \xrightarrow[\zeta_1 \to \infty]{} 0, \end{aligned}\right\} \tag{44}$$

where $\zeta_1 = \psi(\tau)\zeta$.

Allowing for the condition that problem (43) is solvable, i.e.

$$\frac{\partial \alpha_{-1}}{\partial \tau} = 0, \tag{45}$$

we immediately find

$$\overset{1}{V}{}^{\alpha}_{2} = 0, \quad \overset{1}{V}{}^{\nu}_{2} = H_1(\tau, \eta)\, v(\zeta_1 - \xi_0), \tag{46}$$

where $H_1(\tau, \eta)$ is an arbitrary function of the same form as H_0.

We introduce the differential operator

$$L = \left[\frac{\partial^2}{\partial \eta^2} - \left(\frac{\partial \alpha_0}{\partial \eta}\right)^2 + 2i\, \frac{\partial \alpha_0}{\partial \eta}\, \frac{\partial}{\partial \eta} + i\, \frac{\partial^2 \alpha_0}{\partial \eta^2}\right] + 2i\, \frac{\partial}{\partial \tau} - K(\tau)\, \eta^2 - 2\, \frac{\partial \alpha_0}{\partial \tau} + i\, \frac{\partial}{\partial \tau} \ln\left(\sqrt{g}\, \mu G^{\tau\tau}\right), \tag{47}$$

in which

$$K(\tau) = \frac{1}{2}\frac{\partial^2}{\partial \alpha^2}(b^2 G^{\tau\tau})\big|_{\alpha=0,\ \nu=0}. \tag{48}$$

Then for $\overset{2}{V}{}^{\nu}_{2}$ and $\overset{2}{V}{}^{\alpha}_{2}$ we will have the following Sturm-Liouville problems:

$$\left.\begin{aligned} &\frac{\partial^2 \overset{2}{V}{}^{\nu}_{2}}{\partial \zeta_1^2} - (\zeta_1 - \xi_0)\overset{2}{V}{}^{\nu}_{2} = -\frac{1}{b^2\psi^2} L \overset{0}{V}{}^{\nu}_{2}, \\ &\overset{2}{V}{}^{\nu}_{2}\big|_{\zeta_1=0} = -4Mb^2\psi \frac{\partial \overset{0}{V}{}^{\nu}_{2}}{\partial \zeta_1}\bigg|_{\zeta_1=0}, \quad \overset{2}{V}{}^{\nu}_{2} \underset{\zeta_1\to\infty}{\to} 0; \end{aligned}\right\} \tag{49}$$

$$\left.\begin{aligned} &\frac{\partial^2 \overset{2}{V}{}^{\alpha}_{2}}{\partial \zeta_1^2} - (\zeta_1 - \xi_0)\overset{2}{V}{}^{\alpha}_{2} = f(\tau, \eta, \zeta_1), \\ &\frac{\partial \overset{2}{V}{}^{\alpha}_{2}}{\partial \zeta_1}\bigg|_{\zeta_1=0} = 0, \quad \overset{2}{V}{}^{\alpha}_{2} \underset{\zeta_1\to\infty}{\to} 0, \end{aligned}\right\} \tag{50}$$

where

$$f(\tau, \eta, \zeta_1) = \frac{2ib_{\tau\alpha}}{b^4\psi^2}\overset{0}{V}{}^{\nu}_{2}(\tau, \eta, \zeta_1).$$

Problem (49) is a particular case of the more general Sturm-Liouville problem

$$\left.\begin{aligned} &\frac{\partial^2 V^{\nu}}{\partial \zeta_1^2} - (\zeta_1 - \xi_0) V^{\nu} = F(\tau, \eta, \zeta_1), \\ &V^{\nu}\big|_{\zeta_1=0} = \beta(\tau, \eta), \quad V^{\nu} \underset{\zeta_1\to\infty}{\to} 0. \end{aligned}\right\} \tag{51}$$

Let us consider the solution of this problem. The necessary and sufficient condition for it to be solvable can be written as follows:

$$\int_0^\infty F(\tau, \eta, x)\, v(x - \xi_0)\, dx = \beta \cdot v'(-\xi_0). \tag{52}$$

Obviously the unknown function V^ν is defined to within a term $H(\tau,\eta)v$, where $H = H(\tau,\eta)$ is an arbitrary function.

We construct the Green function

$$G^{\nu}(\zeta_1, x) = \left|\begin{aligned} &\frac{v(x-\xi_0)\,[u(\zeta_1-\xi_0)\,v'(-\xi_0) - v(\zeta_1-\xi_0)\,u'(-\xi_0)]}{W[u,v]\,v'(-\xi_0)} && \text{for } \zeta_1 \leqslant x, \\ &\frac{v(\zeta_1-\xi_0)\,[u(x-\xi_0)\,v'(-\xi_0) - v(x-\xi_0)\,u'(-\xi_0)]}{W[u,v]\cdot v'(-\xi_0)} && \text{for } \zeta_1 \geqslant x. \end{aligned}\right\} \tag{53}$$

The functions $u = \text{Re} w_1$ and $v = \text{Re} w_1$ in (53) are linearly independent solutions of the Airy equation $w_1'' - tw_1 = 0$[5], $W[u,v]$ is the Wronskian of the linearly independent solutions, and

$$W[u, v] = \frac{1}{2}. \tag{54}$$

The final solution of problem (51) is found by the standard method [7]:

$$V^{\nu} = H(\tau, \eta)\, v(\zeta_1 - \xi_0) + \int_0^{\infty} G^{\nu}(\zeta_1, x)\, F(\tau, \eta, x)\, dx. \tag{55}$$

Let us, reconsider expressions (49). Allowing for (54) and assuming that the solvability condition (52) holds, we find the solution of problem (49) in the form

$$\overset{2}{V}{}_2^{\nu} = H_2(\tau, \eta)\, v(\zeta_1 - \xi_0) + \int_0^{\infty} G^{\nu}(\zeta_1, x)\, F(\tau, \eta, x)\, dx, \tag{56}$$

where

$$F(\tau, \eta, \zeta_1) = -\frac{1}{b^2\psi^2} L\overset{0}{V}{}_2^{\nu} = -\frac{1}{b^2\psi^2}\left[LH_0 \cdot v + 2i\frac{\psi'}{\psi} H_0 v' \zeta_1\right], \tag{57}$$

$$H_2(\tau, \eta) = A_2(\tau, \eta)\, D_q + C_2(\tau, \eta)\, D_q'. \tag{58}$$

Now consider condition (52) for the solvability of problem (49). After some operations it may be written as follows:

$$LH_0 - 4Mb^4\psi^3 H_0 - i\frac{\psi'}{\psi} H_0 = 0. \tag{59}$$

It should be noted that in deriving (59) we used the equations

$$\int_0^{\infty} v^2(x - \xi_0)\, dx = [v'(-\xi_0)]^2, \qquad \int_0^{\infty} v \cdot v' \cdot x\, dx = -\frac{[v'(-\xi_0)]^2}{2},$$

which are given in Dorodnicyn's article [8].

The function $H_0(\tau, \eta)$ is specified by formula (40); by assumption, A_0 and C_0 are polynomials in η with coefficients which depend on the parameter τ, while D_q and D_q' are linearly independent functions. Equation (59) is satisfied when the coefficients for D_q' and D_q are separately equal to zero, i.e.

$$\frac{\partial^2 A_0}{\partial\eta^2}+\frac{2}{\gamma^2}\left[\frac{\eta^2}{2\gamma^2}-\left(q+\frac{1}{2}\right)\right]A_0+2i\frac{\partial A_0}{\partial\tau}+i\frac{\partial^2\alpha_0}{\partial\eta^2}A_0$$

$$+\frac{2\sqrt{2}}{\gamma}\left[\frac{\eta^2}{2\gamma^2}-\left(q+\frac{1}{2}\right)\right]\frac{\partial C_0}{\partial\eta}+\sqrt{2}\,C_0\frac{\eta}{\gamma^3}-\left(\frac{\partial\alpha_0}{\partial\eta}\right)^2A_0$$

$$+2i\frac{\partial\alpha_0}{\partial\eta}\left\{\frac{\partial A_0}{\partial\eta}+\sqrt{2}\frac{C_0}{\gamma}\left[\frac{\eta^2}{2\gamma^2}-\left(q+\frac{1}{2}\right)\right]\right\}-2\frac{\partial\alpha_0}{\partial\tau}A_0$$

$$-K(\tau)\,\eta^2A_0+i\frac{\partial}{\partial\tau}\ln\left(\sqrt{g}\,\mu\,G^{\tau\tau}\right)\cdot A_0+4Mb^4\psi^3A_0-i\frac{\psi'}{\psi}A_0=0, \tag{60}$$

$$\frac{\partial^2 C_0}{\partial\eta}+\frac{2\sqrt{2}}{\gamma}\frac{\partial A_0}{\partial\eta}+\frac{2}{\gamma^2}\left[\frac{\eta^2}{2\gamma^2}-\left(q+\frac{1}{2}\right)\right]C_0-\left(\frac{\partial\alpha_0}{\partial\eta}\right)^2C_0$$

$$+2i\frac{\partial\alpha_0}{\partial\eta}\left[A_0\frac{\sqrt{2}}{\gamma}+\frac{\partial C_0}{\partial\eta}\right]+i\frac{\partial^2\alpha_0}{\partial\eta^2}C_0+2i\frac{\partial C_0}{\partial\tau}-i\frac{\psi'}{\psi}C_0$$

$$-K(\tau)\,\eta^2C_0+2i\frac{\sqrt{2}}{\gamma^2}(-\gamma')\,A_0+i\frac{\partial}{\partial\tau}\ln\left(\sqrt{g}\,\mu G^{\tau\tau}\right)C_0$$

$$-2\frac{\partial\alpha_0}{\partial\tau}C_0+4Mb^4\psi^3C_0=0. \tag{61}$$

Equating the coefficients for different powers of η to zero and performing some operations, we obtain

$$\frac{\partial A_0}{\partial\eta}=0,\quad \frac{\partial\alpha_0}{\partial\eta}=\eta\frac{\gamma'}{\gamma},\quad C_0=0,\quad \gamma(\tau)=\frac{1}{\chi(\tau)}. \tag{62}$$

Here

$$\alpha_0(\tau,\ \eta)=\alpha_{00}(\tau)+\alpha_{01}(\tau)\,\eta+\alpha_{02}(\tau)\,\eta^2, \tag{63}$$

where

$$\alpha_{01}(\tau)=0,\quad \alpha_{02}(\tau)=\frac{1}{2}\frac{\gamma'}{\gamma}. \tag{64}$$

The functions $A_0(\tau)$ and $\alpha_{00}(\tau)$ are as follows:

$$A_0=A_0(\tau)=\text{const}\cdot\sqrt{\frac{G_{\tau\tau}\psi}{\sqrt{g}\,\mu\gamma}}, \tag{65}$$

$$\alpha_{00}(\tau)=\frac{1}{2}\int_0^\tau\left(-\frac{2q+1}{\gamma^2}+4Mb^4\psi^3\right)d\tau, \tag{66}$$

while the function $\gamma(\tau)$ satisfies the following equation:

$$\gamma''+K(\tau)\,\gamma=\frac{1}{\gamma^3}. \tag{67}$$

The solution of the last equation is given by formula (29) of [3]. Henceforth $\gamma(\tau)$ will be assumed to be known.

Thus the solvability condition of problem (49) for the function $\overset{2}{V}{}_2^{\nu}$ has made it possible to find the functions $H_0(\tau,\eta)$ and $\overset{0}{V}{}_2^{\nu}$ correct to an arbitrary constant factor:

$$\left.\begin{aligned} \overset{0}{V}{}_0^{\nu} &= \text{const}\cdot\sqrt{\frac{G_{\tau\tau}\psi}{\sqrt{g}\,\mu\gamma}}\exp\left\{-\frac{i}{2}\left[-\frac{\gamma'}{\gamma}\eta^2+\int_0^{\tau}\left(\frac{2q+1}{\gamma^2}-4Mb^4\psi^3\right)d\tau\right]\right\}\times \\ &\times v(\zeta_1-\xi_0)\,D_q\left(\sqrt{2}\frac{\eta}{\gamma}\right), \\ \xi_0 &= x_p,\quad v(-\xi_0)=0,\quad p=1,\,2,\,\ldots,\quad q=0,\,1,\,2,\,\ldots \end{aligned}\right\} \tag{68}$$

Let us return to problem (50). Its solution may be written in the form

$$\overset{2}{V}{}_2^{\alpha}=\int_0^{\infty}G^{\alpha}(\zeta_1,\,x)\,f(\tau,\,\eta,\,x)\,dx, \tag{69}$$

where $G^{\alpha}(\zeta_1,x)$ denotes the Green function of the corresponding homogeneous problem; it is

$$G^{\alpha}(\zeta_1,\,x)=\left|\begin{array}{ll} \dfrac{v(x-\xi_0)\,[u(\zeta_1-\xi_0)\,v'(-\xi_0)-v(\zeta_1-\xi_0)\,u'(-\xi_0)]}{W[u,\,v]\,v'(-\xi_0)} & \text{for } \zeta_1\leqslant x, \\ \dfrac{v(\zeta_1-\xi_0)\,[u(x-\xi_0)\,v'(-\xi_0)-v(x-\xi_0)\,u'(-\xi_0)]}{W[u,\,v]\,v'(-\xi_0)} & \text{for } \zeta_1\geqslant x. \end{array}\right\} \tag{70}$$

In this formula, as before, u and v denote linearly independent solutions of the Airy equation, $W[u,v]$ is the Wronskian (54), and the function f is given by

$$f(\tau,\,\eta,\,x)=\frac{2ib_{\tau\alpha}}{\psi^2b^4}\cdot\overset{0}{V}{}_2^{\nu}(\tau,\,\eta,\,x).$$

The adjoint homogeneous equation which corresponds to the Sturm-Liouville problem (50),

$$\frac{\partial^2V}{\partial\zeta_1^2}-(\zeta_1-\xi_0)\,V=0,$$

has only the trivial solution corresponding to the conditions

$$\left.\frac{\partial V}{\partial\zeta_1}\right|_{\zeta_1=0}=0,\quad V\underset{\zeta_1\to\infty}{\to}0.$$

Therefore problem (50) is uniquely solvable for any right side $f(\tau,\eta,x)$, and its solution has the form (69)–(70).

The basic results obtained in this stage are as follows. The functions $\overset{0}{V}{}_2^{\nu}$ and $\overset{0}{V}{}_2^{\alpha}$ are defined by (68)–(70) correct to an arbitrary constant factor. The function V_2^{ν} (see formulas (56)–(58)) has the term $H_2\cdot v$ where H_2 will be computed later from the condition that the Sturm-Liouville problem of the form (51) be solvable for the function $\overset{4}{V}{}_2^{\nu}$. We note that all the coefficients α_k have been determined.

Let us now determine the coefficients of $\overset{k}{V}{}_2^\nu$ and $\overset{k}{V}{}_2^\alpha$, $k \geqslant 3$, and the functions $H_k(\tau,\eta)$, $k \geqslant 1$:

$$H_k(\tau, \eta) = A_k(\tau, \eta) D_q + C_k(\tau, \eta) D'_q. \tag{71}$$

For $\overset{3}{V}{}_2^\nu$ we have

$$\left.\begin{aligned} &\frac{\partial^2 \overset{3}{V}{}_2^\nu}{\partial \zeta_1^2} - (\zeta_1 - \xi_0)\overset{3}{V}{}_2^\nu = F(\tau, \eta, \zeta_1), \\ &\overset{3}{V}{}_2^\nu\Big|_{\zeta_1=0} = -4Mb^2 \frac{\partial \overset{1}{V}{}_2^\nu}{\partial \zeta_1}\Big|_{\zeta_1=0}, \quad \overset{3}{V}{}_2^\nu \xrightarrow[\zeta_1\to\infty]{} 0, \end{aligned}\right\} \tag{72}$$

where

$$F = -\frac{1}{b^2\psi^2}\Big[LH_1 \cdot v + 2iv'\zeta_1 H_1 \frac{\psi'}{\psi} + \frac{4ib_{\tau\alpha}}{\psi g_{\tau\tau}}\zeta_1 \cdot \frac{\partial H_0}{\partial \eta} v - \frac{4b_{\tau\alpha}}{\psi g_{\tau\tau}}\zeta_1$$
$$\times \eta \frac{\gamma'}{\gamma} H_0 \cdot v + \eta \frac{\partial b^2}{\partial \alpha}\frac{1}{\psi^2}(\zeta_1 - \zeta_0) H_0 v - \frac{1}{2}\eta\zeta_1 \frac{1}{\psi} \cdot \frac{\partial^2}{\partial\alpha\partial\nu}(G^{\tau\tau}b^2) \cdot H_0 v\Big].$$

The solution of the Sturm-Liouville problem (72) is given by the formula

$$\overset{3}{V}{}_2^\nu = (A_3 D_q + C_3 D'_q)\, v + \int_0^\infty G^\nu(\zeta_1, x) F(\tau, \eta, x)\, dx, \tag{73}$$

where the Green function $G^\nu(\zeta_1, x)$ is specified by (53).

The solvability condition for problem (72) is

$$LH_1 - 4Mb^4\psi^3 H_1 - i\frac{\psi'}{\psi}H_1 - \frac{1}{2}\eta\frac{1}{\psi}\frac{\partial^2}{\partial\alpha\partial\nu}(b^2G^{\tau\tau}) H_0\xi_0$$
$$+ \frac{4b_{\tau\alpha}}{g_{\tau\tau}\psi}\Big(i\frac{\partial H_0}{\partial\eta}\xi_0 - \eta\cdot\xi_0\frac{\gamma'}{\gamma}H_0\Big) = 0. \tag{74}$$

Equating to zero the coefficients of (74) for the functions D_q and D'_q, and then for different powers of η, we obtain the functions $A_1(\tau,\eta)$ and $C_1(\tau,\eta)$ in the form

$$A_1(\tau, \eta) = A_0(\tau)[A_{10} + A_{11}(\tau)\cdot\eta], \tag{75}$$

$$C_1(\tau, \eta) = A_0(\tau)\cdot\frac{C(\tau)}{\gamma(\tau)}, \tag{76}$$

where

$$A_{10} = \text{const}, \quad A_{11}(\tau) = -\Big[\frac{2ib_{\tau\alpha}\xi_0}{\psi g_{\tau\tau}} + \frac{\sqrt{2}}{\gamma^{-1}}\frac{\partial}{\partial\tau}\Big(\frac{c}{\gamma}\Big)\Big], \tag{77}$$

and the function $C(\tau)$ is a solution of the equation

$$C''(\tau) + K(\tau)C(\tau) = f(\tau), \tag{78}$$

$$f(\tau) = \frac{\xi_0}{\sqrt{2}}\Big[\frac{4\partial}{\partial\tau}\Big(\frac{b_{\tau\alpha}}{g_{\tau\tau}\psi}\Big) - \frac{1}{2\psi}\frac{\partial^2}{\partial\alpha\partial\nu}(b^2G^{\tau\tau})\Big]. \tag{79}$$

Thus formulas (46) and (73) define the coefficient $\overset{1}{V}{}_2^\nu$ completely, and they define $\overset{3}{V}{}_2^\nu$ correct to a term $H_3 v$. The function H_3 will be determined later from the condition that the corresponding Sturm-Liouville problem be solvable for the function $\overset{5}{V}{}_2^\nu$, etc.

The function $\overset{3}{V}{}_2^\alpha$ satisfies the following problem:

$$\left.\begin{aligned} &\frac{\partial^2 \overset{3}{V}{}_2^\alpha}{\partial \zeta_1^2} - (\zeta_1 - \xi_0)\overset{3}{V}{}_2^\alpha = \frac{2ib_{\tau\alpha}}{b^4\psi^2}\overset{1}{V}{}_2^\nu, \\ &\left.\frac{\partial \overset{3}{V}{}_2^\alpha}{\partial \zeta_1}\right|_{\zeta_1=0} = -\frac{1}{b^2\psi}\left.\frac{\partial \overset{2}{V}{}_2^\nu}{\partial \eta}\right|_{\zeta_1=0}, \quad \overset{3}{V}{}_2^\alpha \xrightarrow[\zeta_1\to\infty]{} 0. \end{aligned}\right\} \tag{80}$$

Introducing the Green function $G^\alpha(\zeta_1, x)$ from (70), we obtain the desired function in the form

$$\overset{3}{V}{}_2^\alpha = \int_0^\infty G^\alpha(\zeta_1, x)\frac{2ib_{\tau\alpha}}{b^4\psi^2}\overset{1}{V}{}_2^\nu(\tau, \eta, x)\,dx - \frac{1}{b^2\psi}\left.\frac{\partial \overset{2}{V}{}_2^\nu}{\partial \eta}\right|_{\zeta_1=0} \cdot \frac{v(\zeta_1 - \xi_0)}{v'(-\xi_0)}. \tag{81}$$

We note that $\partial \overset{2}{V}{}_2^\nu/\partial\eta$ is known for $\zeta_1 = 0$ at this stage (see (56)–(68)).

In the higher-order approximations, the functions $\overset{k}{V}{}_2^\nu$, $k = 4, 5, \ldots$, will be determined from equations of the form

$$\left.\begin{aligned} &\frac{\partial^2 \overset{k}{V}{}_2^\nu}{\partial \zeta_1^2} - (\zeta_1 - \xi_0)\overset{k}{V}{}_2^\nu = -\frac{1}{b^2\psi^2}\left[L\overset{k-2}{V}{}_2^\nu + f^{(k)}\right], \\ &\left.\overset{k}{V}{}_2^\nu\right|_{\zeta_1=0} = \varphi_k, \quad \overset{k}{V}{}_2^\nu \xrightarrow[\zeta_1\to\infty]{} 0, \end{aligned}\right\} \tag{82}$$

while the functions $\overset{k}{V}{}_2^\alpha$ will be given by

$$\left.\begin{aligned} &\frac{\partial^2 \overset{k}{V}{}_2^\alpha}{\partial \zeta_1^2} - (\zeta_1 - \xi_0)\overset{k}{V}{}_2^\alpha = -\frac{1}{b^2\psi^2}g^{(k)}, \\ &\left.\frac{\partial \overset{k}{V}{}_2^\alpha}{\partial \zeta_1}\right|_{\zeta_1=0} = \psi_k, \quad \overset{k}{V}{}_2^\alpha \xrightarrow[\zeta_1\to\infty]{} 0, \end{aligned}\right\} \tag{83}$$

where

$$f^{(k)} = f^{(k)}\left(\tau, \eta, \zeta, \overset{0}{V}{}_2^\nu, \overset{1}{V}{}_2^\nu, \ldots, \overset{k-3}{V}{}_b^\nu, \overset{2}{V}{}_2^\alpha, \overset{3}{V}{}_2^\alpha, \ldots, \overset{k-2}{V}{}_2^\alpha\right),$$

$$g^{(k)} = g^{(k)}\left(\tau, \eta, \zeta, \overset{0}{V}{}_2^\nu, \overset{1}{V}{}_2^\nu, \ldots, \overset{k-3}{V}{}_2^\nu, \overset{2}{V}{}_2^\alpha, \overset{3}{V}{}_2^\alpha, \ldots, \overset{k-2}{V}{}_2^\alpha\right),$$

$$\varphi_k = \varphi_k(\tau, \eta), \quad \psi_k = \psi_k(\tau, \eta).$$

In solving problems (82) and (83) successively for each k the functions $f^{(k)}$, $g^{(k)}$, φ_k and ψ_k are known, and the desired coefficients are

$$\left.\begin{aligned} \overset{k}{V}{}_2^{\nu} &= H_k(\tau,\ \eta)\, v - \int_0^{\infty} G^{\nu}(\zeta_1,\ x) \frac{1}{b^2\psi^2}\left[L\overset{k-2}{V}{}_2^{\nu} + f^{(k)}\right] dx, \\ \overset{k}{V}{}_2^{\alpha} &= \frac{v(\zeta_1 - \xi_0)}{v'(-\xi_0)}\psi_k + \int_0^{\infty} G^{\alpha}(\zeta_1,\ x)\, g^{(k)} dx. \end{aligned}\right\} \qquad (84)$$

The first of these formulas contains the arbitrary function $H_k(\tau, \eta)$, which will be found from the condition that the problem (82) be solvable for the coefficient $\overset{k+2}{V}{}_2^{\nu}$.

Any coefficient $\overset{k}{V}{}_2^{\nu}$ obtained in the form (84) may be reduced to the form (16) by using the properties of Airy functions and the representations of integrals of products of Airy functions $u(x)$ and $v(x)$ and of finite powers of the variable x, given in [8].

§5. In the preceding sections we have considered the construction of asymptotic series in powers of the small parameter $\omega^{-1/6}$ for the components of the displacement vector. We will now offer some conclusions and will discuss the physical meaning of our solution.

1. The set of high-frequency steady-state waves close to the surface depends on two integer valued parameters $p = 1, 2, \ldots$ and $q = 0, 1, \ldots$. Each wave consists of a longitudinal component $\mathbf{u}_a$ and a transverse component $\mathbf{u}_b$; these propagate along S with the same phase velocity:

$$v_{\phi}^{(p,\ q)} = b(\tau,\ 0,\ 0) + \omega^{-2/3}\frac{\xi_0\psi^2 b^3}{2} + \omega^{-1}\frac{b}{2}\left(\frac{2q+1}{\gamma^2} - 4Mb^4\psi^3\right) + O(\omega^{-4/3}).$$

The sum $\mathbf{u} = \mathbf{u}_a + \mathbf{u}_b$ satisfies the condition that there are no stresses on the boundary S.

With increasing distance from the boundary, the longitudinal component $\mathbf{u}_a$ decreases as $\exp(-\omega\nu M)$, where

$$M = \sqrt{\frac{1}{g_{\tau\tau}} - \frac{1}{a^2}}\,\Bigg|_L.$$

When condition (6) holds, the transverse component $\mathbf{u}_b$ is concentrated near the boundary S in a layer of order $\omega^{-2/3}$ taken in terms of ν, forming a "whispering gallery" effect. As we go from this layer into the interior of Ω, we find that the transverse component tends exponentially to zero in the way that $v(x)$ decreases as $x \to \infty$. The depth of penetration of the longitudinal wave is substantially less than that of the transverse wave.

In view of the factors D_q and D'_q, all the waves are concentrated near L in a band of order $\omega^{-1/2}$ taken in terms of α and decrease as $|\alpha|$ subsequently increases.

2. To investigate the qualitative nature of our solutions, we will give their final form, restricting ourselves to the first nonzero coefficients of the expansions of the components of $\mathbf{u}_a$ and $\mathbf{u}_b$. The displacement vector

$$\mathbf{u} = \mathbf{u}_a + \mathbf{u}_b \qquad (85)$$

for $\nu > 0$ has the form

$$\mathbf{u}_a \sim \exp[-i\omega(t-\tau) + i\omega^{1/3}\alpha_{-2} + i\alpha_0] \exp(-\omega\nu M)\, \omega^{-1/3} \left(\overset{2}{V}{}_1^{\tau},\ \omega^{-1/2},\ \overset{5}{V}{}_1^{\alpha},\ \overset{2}{V}{}_1^{\nu} \right), \quad (86)$$

$$\mathbf{u}_b \sim \exp[-i\omega(t-\tau) + i\omega^{1/3}\alpha_{-2} + i\alpha_0] \left(\overset{2}{V}{}_2^{\tau} \omega^{-1/3},\ \omega^{-1/3} \overset{2}{V}{}_2^{\alpha},\ \overset{0}{V}{}_2^{\nu} \right), \quad (87)$$

where α_{-2}, α_0, $\overset{2}{V}{}_1^{\tau}$, $\overset{5}{V}{}_1^{\alpha}$, $\overset{2}{V}{}_1^{\nu}$, $\overset{2}{V}{}_2^{\tau}$, $\overset{2}{V}{}_2^{\alpha}$, $\overset{0}{V}{}_2^{\nu}$ are known functions.

Inside Ω the general solution $\mathbf{u}$ is basically characterized by the vector $\mathbf{u}_b$, since $|\mathbf{u}_a|$ is asymptotically a quantity of lower order than $|\mathbf{u}_b|$. The isolation of $\mathbf{u}_a$ from the general solution for $\nu \neq 0$ is provisional, since the components $|V_2^{\tau}|$ and $|V_2^{\alpha}|$ have the same order of smallness as $|\mathbf{u}_a|$.

The vector $\mathbf{u}_b$ defines waves concentrated near L with respect to α which oscillate in the layer $0(\omega^{-2/3})$ taken with respect to ν like the eigenfunctions of a "whispering gallery." This property permits us to compare $\mathbf{u}_b$ with a Love wave. In the waves under consideration, however, the particles within the body are displaced primarily in parallel to the normal to S, i.e. perpendicularly to the plane containing the tangents to L and to the wave front.

The vector $\mathbf{u}_b$ contains as a factor the Airy function $v(\psi\zeta - \xi_0)$, where ψ, $\xi_0 > 0$. Hence the depth of penetration of the transverse component of $\mathbf{u}$ is

$$h_p = \omega^{-2/3} \cdot \frac{x_p}{\psi}, \quad p = 1, 2, 3, \ldots,$$

and for $\nu > h_p$ the transverse wave attenuates in the way that the function $v(x)$ decreases as $x \to \infty$.

The vector $\mathbf{u}_a$ defines surface waves which are concentrated in terms of α near L in the band $0(\omega^{-1/2})$, in which the medium is displaced in parallel to the plane of normal section of S along L (the component V_1^{α} has a greater order of smallness than V_1^{τ} and V_1^{ν}). Like Rayleigh waves, these waves decrease rapidly as we go from the surface into the interior of the body.

Solution (85)–(87) contains $D_q(\sqrt{2}\,\eta/\gamma)$ as a factor, and therefore it is substantially different from zero in the band

$$|\alpha| \leqslant \frac{\sqrt{2q+1}\,\gamma}{\sqrt{\omega}}, \quad q = 0, 1, 2, \ldots, \quad (88)$$

whose width is of order $\omega^{-1/2}$. Outside this band the solution tends to zero as $\eta \to \infty$.

3. On S, i.e. for $\nu = 0$, the longitudinal wave vector $\mathbf{u}_a$ retains the form (86), while for the asymptotic representations of the components of the transverse wave vector $\mathbf{u}_b$ we obtain

$$\mathbf{u}_b \sim \exp[-i\omega(t-\tau) + i\omega^{1/3}\alpha_{-2} + i\alpha_0] \left(\omega^{-1/3} \overset{2}{V}{}_2^{\tau},\ \omega^{-5/6} \overset{5}{V}{}_2^{\alpha},\ \omega^{-1/3} \overset{2}{V}{}_2^{\nu} \right), \quad (89)$$

where the function $\overset{2}{V}{}_2^{\nu}$ is defined by (56).

As for Rayleigh waves, the displacement $\mathbf{u} = \mathbf{u}_a + \mathbf{u}_b$ on S is parallel to the plane of normal section of the surface along L, although the propagation velocity of these waves over the surface is close to the transverse velocity and hence differs from the Rayleigh velocity.

4. By introducing the cartesian coordinate system x, y, z with its center at some fixed point on L, we may show that in the particular case of an elastic subspace which is inhomogeneous with respect to two coordinates, the corresponding formulas of [1] for the components of the displacement vector, depth of penetration, and phase velocity coincide with those obtained above. In comparing the formulas it should be taken into account that the solutions constructed in this paper are concentrated near L with respect to the coordinate α.

We should note that L.M. Brehovskih [9] considered similar waves for cylindrical and spherical boundaries; he called them surface waves with vertical polarization which are confined by the curvature of the boundary.

The author is deeply grateful to his advisor, Professor V.M. Babič, for formulating the problem and for his constant assistance.

BIBLIOGRAPHY

1. I.V. Muhina and I.A. Molotkov, *Propagation of Rayleigh waves in an elastic half-space which is inhomogeneous with respect to two coordinates*, Izv. Akad. Nauk SSSR Ser. Fiz. Zemli 1967, no. 4, 3–8 = Phys. Solid Earth **1967**, 209–211.
2. V.S. Buldyrev, *Interference of short waves in the problem of diffraction on an inhomogeneous cylinder of arbitrary cross section*, Izv. Vysš. Učebn. Zaved. Radiofizika **10** (1967), 699–711.
3. V.M. Babič and V.F. Lazutkin, *The eigenfunctions which are concentrated near a closed geodesic*, Problemy Mat. Fiz., no. 2, Izdat. Leningrad. Univ., Leningrad, 1967, pp. 15–25; English transl., Topics in Math. Phys., no. 2, Consultants Bureau, New York, 1968. MR **38** #2708.
4. V.F. Lazutkin, *Construction of an asymptotic series of eigenfunctions of the "bouncing ball" type*, Trudy Mat. Inst. Steklov. **95** (1968), 106–118 = Proc. Steklov Inst. Math. **95** (1968), 125–140. MR **42** #2194.
5. V.A. Fok, *Tables of Airy functions*, Naučno-Issled. Inst., Inform. Otd. **108** (1946).
6. E.T. Whittaker and G.N. Watson, *A course of modern analysis*, Cambridge Univ. Press, New York, 1927; Russian transl., vol. 2, Fizmatgiz, Moscow, 1963, Chap. 16. MR **31** #2375.
7. V.I. Smirnov, *A course in higher mathematics*. Vol. 4, Fizmatgiz, Moscow, 1958, Chap. 4; English transl. of 3rd ed., Pergamon Press, Oxford; Addison-Wesley, Reading, Mass., 1964. MR **31** #1333.
8. A.A. Dorodnicyn, *Asymptotic laws of distribution of the characteristic values for certain special forms of differential equations of the second order*, Uspehi Mat. Nauk **7** (1952), no. 6 (52), 3–96; English transl., Amer. Math. Soc. Transl. (2) **16** (1960), 1–101. MR **14**, 876; **22** #8161.
9. L.M. Brehovskih, *Surface waves in a solid which are confined by the curvature of the boundary*, Akust. Ž. **13** (1967), 541–555 = Soviet Physics Acoustics **13** (1967), 462–472.

Translated by:
M. Lopez-Morillas

ANALYTIC NATURE OF THE WAVE FIELD IN THE NEIGHBORHOOD OF A CAUSTIC WITH REFLECTION

D. Š. MOGILEVSKIĬ

UDC 517 944/947

Abstract. The article considers the plane problem of the reflection of a caustic from an analytic boundary. It is shown how the wave field in the neighborhood of the caustic may be written in the form of integrals which are analogous to the Poisson integrals for solving the problem for the wave equation.

Figures: 1. Bibliography: 4 items.

Introduction

In this paper we offer a mathematically rigorous treatment of the problem of the analytic nature of the wave field in the neighborhood of a caustic. We will study solutions of the wave equation with variable velocity

$$\frac{1}{c^2(x,\ y)} U_{tt} = \Delta U,$$

which have discontinuities on the characteristic surfaces (wave fronts) $t = \tau(x,y)$, where $\tau(x,y)$ satisfies the eikonal equation

$$(\nabla\tau)^2 = \frac{1}{c^2(x,\ y)}$$

The surfaces of a discontinuity consist of rays (bicharacteristics). When projected on the xy plane the rays are extremals of the variational problem for the Fermat functional

$$I = \int \frac{ds}{c(x,\ y)}.$$

The family of rays covering the surface of a discontinuity may be specified by the vector valued function $\mathbf{X}(\alpha,\tau)$; α characterizes a ray and τ characterizes a point on a ray. A caustic is the envelope of a family of rays. It is defined by the equation $\mathbf{X}_\alpha = 0$. The problem of propagation of discontinuities along a wave front may be solved by the ray method [1]. The solution $U(x,y;t)$ near the front $t = \tau(x,y)$ is sought in the form of a ray series

AMS 1970 *subject classifications* Primary 35L05; Secondary 73D99, 73N99, 86A15.

$$U = \sum_{k=0}^{\infty} f_k(\alpha, \tau)(t-\tau)_+^{\lambda+k}.$$

Here s_+^λ is the standard discontinuity function:

$$s_+^\lambda = \begin{vmatrix} s^\lambda & \text{for} & s > 0, \\ 0 & \text{for} & s \leqslant 0. \end{vmatrix}$$

For $f_k(\alpha, \tau)$ we obtain a recurrent system of ordinary differential equations which is solved with the initial data $f_k(\alpha, 0)$. It is shown that the ray series converges in the neighborhood of the front, except for the caustics, if the series for the initial data converges [2]. The ray series diverges in the neighborhood of the caustic. The analytic nature of the solution in this neighborhood was studied in [3]. The principal singularity is expressed in terms of hypergeometric and Legendre functions. The convergence of ray series in the presence of reflecting and refracting boundaries was studied in [2]; however, the analytic nature of the wave field when a caustic goes to a reflecting or refracting boundary remained unstudied. In this paper this problem is solved by a method similar to that used in [3].

§1. Statement of the problem

First we consider the most elementary problem of the reflection of a caustic.

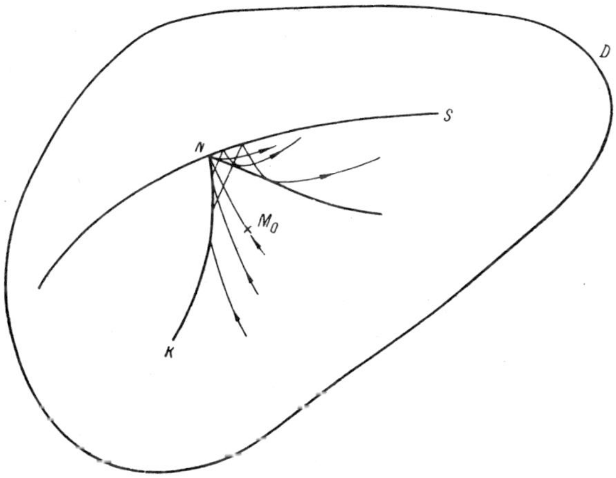

Assume that in region D (see the figure) on the xy plane we are given an analytic function $c(x, y) > 0$. Assume that in the same region we are given the ray field $\mathbf{X}(\alpha, \tau)$ having a caustic K with equation $\mathbf{X}_\alpha = 0$. The vector valued function $\mathbf{X}(\alpha, \tau)$ is assumed to be analytic. The reflecting boundary S is specified by the explicit equation $y = \psi(x)$, where $\psi(x)$ is an analytic function. A ray $\mathbf{X}(\alpha_N, \tau)$ passes through the point N where the caustic K intersects the boundary S. We will assume that the ray α_N is not tangent to S. The point N corresponds to the ray coordinates (α_N, τ_N). Assume that for $\tau < \tau_N$ in the neighborhood of α_N we are given the ray series

$$F(x, y; t) = \sum_{k=0}^{\infty} f_k(\alpha, \tau)(t-\tau)_+^{\lambda+k},$$

which converges in the neighborhood of the front $t = \tau(x,y)$ and satisfies the equation

$$\frac{1}{c^2(x, y)} F_{tt} = \Delta F.$$

We assume $f_k(\alpha, \tau)$ to be analytic in its variables in the region, as is usually assumed for ray series. We also assume that $\mathrm{Re}\lambda > 0$.

Under these conditions we will construct a function $U(x,y;t)$ which satisfies the wave equation $U_{tt}/c^2(x,y) = \Delta U$ and the boundary condition $U|_S = 0$ and is analytic in $(x,y;t)$ in some sphere of radius $\delta > 0$ with its center at the point (N, τ_M), except for the fronts of the incident and reflected waves. In the neighborhood of the incident wave front $U(x,y;t)$ may be written as $U = F + V$, where $V(x,y;t)$ is analytic in this neighborhood. This representation ceases to make sense in the neighborhood of K, where the ray series F diverges.

§2. The Green function

The Green function $G(M,t;M_0,t_0)$ is the solution of the boundary value problem for the wave equation with boundary condition $G|_S = 0$ and initial condition

$$G|_{t=t_0} = 0; \quad G_t|_{t=t_0} = \delta(M - M_0).$$

Here G has discontinuities on the characteristic conoid with vertex at (M_0, t_0) and on the reflected front. This function breaks down into two parts: $G = G_0 + G_1$, where G_0 is the Hadamard kernel and G_1 is the reflected part.

The Hadamard kernel G_0 is the solution of the Cauchy problem with initial data

$$G_0|_{t=t_0} = 0; \quad G_{0t}|_{t=t_0} = \delta(M - M_0).$$

The fact that $c(x,y)$ is analytic implies that there exists an $\varepsilon > 0$ such that for any point $M_0 \in D$ the rays departing from it do not intersect, i.e. they form a field, for

$$\tau(M, M_0) = \int_{M_0}^{M} \frac{ds}{c(x, y)} \leqslant \varepsilon.$$

This means that any characteristic conoid has no caustics for $t - t_0 \leqslant \varepsilon$. In this case the Hadamard kernel $G_0(M,t;M_0,t_0)$ for $t - t_0 \leqslant \varepsilon$ has the following properties:

1) $G_0 = 0$ below a characteristic conoid, i.e. for $t - t_0 < \tau(M, M_0)$;

2) above a characteristic conoid

$$G_0(M, t; M_0, t_0) = \frac{B_0(M, M_0; t - t_0)}{\sqrt{(t - t_0)^2 - \tau^2(M, M_0)}},$$

where $B_0(M, M_0; t - t_0)$ is an analytic function of all its arguments.

It is clear that in the neighborhood of a characteristic conoid, G_0 is a ray series in powers of $(t - t_0 - \tau(M, M_0))_+^{k-1/2}$.

We will require the following theorem about the Green function.

Assume that M_1 is a point on the boundary S, ℓ is a ray departing from M_1 in a direction which is not tangent to the boundary, and M_0 is a point on ℓ such that

$$\tau(M_0, M_1) = \int_{M_0}^{M_1} \frac{ds}{c(x, y)} < \frac{\varepsilon}{2}.$$

Then the reflected part of the Green function $G_1(M, t; M'_0, t_0)$ satisfies the following conditions in a neighborhood $(M_1, t_1; M_0, t_0)$ of radius $\delta_1 > 0$ in all variables: 1) $G_1 = 0$ below the reflected front $t - t_0 = \tau_1(M, M'_0)$, where the function $\tau_1(x, y; x'_0, y'_0)$ is analytic in the neighborhood (M_1, M_0) in all variables, i.e. the reflected front has no caustics; 2) above the reflected front

$$G_1(M, t; M'_0, t_0) = \frac{B_1(M, M'_0; t - t_0)}{\sqrt{t - t_0 - \tau_1(M, M'_0)}},$$

where $B_1(M, M'_0; t - t_0)$ is an analytic function of all its arguments.

We start by proving the first assertion. The family of incident rays is specified by the vector valued function $\mathbf{X}(\alpha, \tau; x'_0, y'_0)$, which is analytic in all its arguments. Here τ is the ray coordinate, (x'_0, y'_0) are the coordinates of the vertex of the conoid M'_0, and α is the angle between the ray direction in M'_0 and some fixed direction. It is understood that $\mathbf{X}(\alpha, \tau; x'_0, y'_0)$ denotes the ray field of the Green function, i.e. its meaning is different from that of the ray field $\mathbf{X}(\alpha, \tau)$ of the initial problem in §1. Here $\mathbf{X}(\alpha, \tau; x'_0, y'_0)$ is analytic as a consequence of the fact that the rays are extrema of the Fermat functional and satisfy the Euler equations with analytic coefficients. Now we need only refer to the theorem on the analyticity of solutions of ordinary differential equations with analytic coefficients with respect to the initial data [4]. The equation of the reflecting boundary is $y = \psi(x)$, where $\psi(x)$ is an analytic function. Substituting the functions of the ray field $\mathbf{X}(\alpha, \tau; x'_0, y'_0)$ into the equation for the boundary, we obtain its equation in ray coordinates:

$$y(\alpha, \tau; x'_0, y'_0) = \psi(x(\alpha, \tau; x'_0, y'_0)).$$

This equation may be solved with respect to τ by the theorem on implicit functions $\tau = \tau'(\alpha, x'_0, y'_0)$. Here $\tau'(\alpha, x'_0, y'_0)$ is analytic in all its arguments. The condition for the equation to be solvable coincides with the condition that a ray be not tangent to the boundary:

$$\frac{\partial y}{\partial \tau} - \psi' \frac{\partial x}{\partial \tau} \neq 0.$$

Now consider the family of reflected rays $\mathbf{X}_1(\tau_1, \alpha_1; x_1', y_1', \tau_1')$. This vector valued function is analytic in all its arguments; τ_1 is a ray parameter and all the remaining variables are initial data. They have the following meaning:

$$\tau_1' = \tau'(\alpha, x_0', y_0'),$$
$$x_1' = x(\alpha, x_0', y_0'; \tau'(\alpha, x_0', y_0')),$$
$$y_1' = y(\alpha, x_0', y_0'; \tau'(\alpha, x_0', y_0')),$$

i.e. τ_1 and α_1 are specified at points of the reflecting boundary (x_1', y_1'). Here α_1 is the angle of a reflected ray at a point of reflection with fixed direction. The function $\alpha_1(\alpha, x_0', y_0')$ is analytic in all variables, since the family of incident rays and the reflecting boundary are analytic. The result is that $\mathbf{X}_1(\alpha, \tau_1; x_0', y_0')$ is analytic in all its arguments (superposition of analytic functions). Thus the family of reflected rays $\mathbf{X}_1(\alpha, \tau_a; x_0', y_0')$ is analytic in all variables. Moreover, $\mathbf{X}_{1\alpha} \neq 0$ at the point (M_1, M_0), since the reflected rays of the Green function $G(M_1, M_0, t - t_0)$ do not have a caustic at the reflection point M_1. The fact that $\mathbf{X}_1(\tau_1, \alpha; x_0', y_0')$ is analytic implies that there exists a neighborhood (M_1, M_0) of radius $\delta_2 > 0$ in all variables in which $\mathbf{X}_{1\alpha} \neq 0$, i.e. there is no caustic. The fact that $G_1 = 0$ below the reflected front results from the uniqueness theorem for the wave equation.

Now we will deal with the second assertion. It is a consequence of the theorem on convergence of ray series in the presence of reflecting boundaries [2]. Admittedly, we require a somewhat more general assertion than the one proved in [2]. It is formulated as follows. If $\tau(x, y; \lambda_1 \ldots \lambda_n)$ is a solution of the eikonal equation which is an analytic function of the parameters $\lambda_1 \ldots \lambda_n$ in some region E of their variation and the ray series $\sum_{k=0}^{\infty} f_k(\alpha, \tau; \lambda_1 \ldots \lambda_n)(t - \tau)_+^{\lambda+k}$ converges for $\tau = \tau_0$ and $(\lambda_1 \ldots \lambda_n)$ in E, then the ray series converges in the neighborhood of the front to a function which is analytic in all arguments $(\alpha, \tau, t; \lambda_1 \ldots \lambda_n)$, where $(\lambda_1 \ldots \lambda_n) \in E$, and (α, τ) up to the caustics. This theorem differs from that proved in [2] by the presence of the parameters $\lambda_1 \ldots \lambda_n$. The proof differs in that there is a check on the uniformity of all estimates in the parameters λ_i. A uniform estimate immediately implies uniform convergence and analyticity of the sum in the λ_i. For us the parameters are the coordinates (x_0', y_0') of the vertex M_0'. They vary in the neighborhood M_0 of radius δ_2 with respect to both variables. In this case there are no caustics in the neighborhood of δ_2, and the theorem we have formulated may be applied. The initial data for the reflected part of the Green function G_1 are analytic in view of the analyticity of $B_0(M, M_0'; t - t_0)$ and of the reflecting boundary S. In accordance with the theorem, the ray series for G_1 will converge in some neighborhood of the reflected front to a function which is analytic in all arguments, QED.

§3. Integral representation of a singularity

The ray α_N passes through the point N. We select the point M_0 on the ray from the condition

$$\int_{M_0}^{N} \frac{dS}{c(x, y)} < \frac{\varepsilon}{2}.$$

Assume that M_0 has the ray coordinates (α_N, τ_0). The given ray series

$$\sum_{k=0}^{\infty} f_k(\alpha, \tau)(t - \tau)_+^{\lambda+k}$$

converges in the neighborhood of the front. Assume that $\delta_3 > 0$ is such that the series converges in a circle of radius δ_3 on the plane $t = \tau_0$ with center at M_0. We take $\delta_4 = \min(\delta_3, \delta_1)$ and consider the integral

$$U(x, y; t) = \int_R [F(x_0', y_0') G_t(x, y; x_0', y_0'; t - t_0) + F_t(x_0', y_0') G(x, y; x_0', y_0'; t - t_0)]\, dx_0' dy_0',$$

Here R is a circle of radius δ_4 on the plane $t = \tau_0$ with center at M_0, and $G(M, M_0'; t - t_0)$ is the Green function.

We will show that the $U(x,y;t)$ we have constructed satisfies the requirements of §2. Indeed, it satisfies the wave equation and the boundary condition, since G possesses these properties.

We are interested in the analytic properties of $U(x,y;t)$ in the neighborhood of the point (N, τ_N). We write U as the sum $U = U_0 + U_1$, where

$$U_0 = \int_R (FG_{0t} + F_t G_0)\, dx_0' dy_0',$$

$$U_1 = \int_R (FG_{1t} + F_t G_1)\, dx_0' dy_0'.$$

In accordance with [3], $U_0(x,y;t)$ is analytic everywhere in the neighborhood of (N, τ_N) except for the front of the incident wave. This is a consequence of the properties of the Hadamard kernel G_0. The integral for U_0 may have a singularity only at those points M whose characteristic conoid is tangent to the initial front [3]. It was also shown in [3] that $U_0 = F + V_0$, where V_0 is analytic in the neighborhood of the incident wave front. We should stipulate, however, that V_0 is analytic only in that neighborhood of (N, τ_N) in which the singularities from the isolation of the circle of radius δ_4 on the plane $t = \tau_0$ have no effect. It is important that such a neighborhood always exists [3].

Let us now consider $U_1(x,y;t)$. In the sphere of radius δ_1 with center (N, τ_N), G_1 has properties which are exactly analogous to those of G_0. It follows that $U_1(x,y;t)$ is analytic in the neighborhood of (N, τ_N) except for the reflected front. Like U_0, a singularity in the integral for U_1 may appear only at those points M whose reflected conoid characteristic is tangent to the initial front:

$$U_1 = F_1 + V_1,$$

where F_1 is the reflected ray series and V_1 is analytic in the neighborhood of the reflected front (with the same stipulation as for V_0).

Thus $U(x,y;t)$ is analytic in some neighborhood of (N, τ_N) except for the incident and reflected wave fronts where there are ray discontinuities, QED. If N is a nonsingular point of the caustic of the incident wave, then it is also nonsingular for the reflected wave, and $U_1(x,y;t)$ may also be expressed in the neighborhood of (N, τ_N) in terms of hypergeometric functions or Legendre functions, as was done in [3] for $U_0(x,y;t)$.

§4. Remarks on generalization

The assertion proved in the preceding section is based on theorems concerning ray series [1–3] and on the properties of the fundamental solution of the Cauchy problem (the Hadamard kernel G_0), and remains valid in all cases in which these theorems are applicable. In particular, the boundary condition may be replaced by any homogeneous reflection condition

$$\left.\frac{\partial U}{\partial n}\right|_S = 0; \quad \left.\frac{\partial U}{\partial n} + \sigma U\right|_S = 0,$$

where σ is an analytic function on the boundary S. A refracting boundary may be considered instead of a reflecting one. The condition $\mathrm{Re}\lambda > 0$ is not essential, since ray series are defined for any λ, and the integral

$$\int (FG_t + F_t G)\, dx_0' dy_0'$$

may be understood in the sense of generalized functions. The theorems concerning ray series and the requisite properties of the Hadamard kernel are independent of the number of spatial variables. Therefore the basic assertion holds for any number of spatial variables, although the caustic manifolds may be extremely complex. At the same time, analyticity must be observed at every single point (coefficients of the equation, initial data, boundary), since the convergence of the ray series is contingent upon it. Also essential is the requirement that a ray not be tangent to the boundary (incidence at the angle of total internal reflection is also inadmissible in the case of refraction), since in this case the ray method fails. However, the requirement of analyticity may be replaced by one of sufficient smoothness if the formulation of the theorem is appropriately modified. Then $F(x,y;t)$ should be understood as the finite sum $\sum_{k=0}^{n} f_k(\alpha,\tau)(t-\tau)_+^{\lambda+k}$, and we may only assert that V will be sufficiently smooth; the degree of smoothness will depend on the differential properties of $c(x,y)$, of the boundary, and of the initial data.

BIBLIOGRAPHY

1. A.S. Alekseev, V.M. Babič and B. Ja. Gel′ činskiĭ, *The ray method of calculating the intensity of wave fronts*, Problems in the Dynamic Theory of the Propagation of Seismic Waves, no. 5, Izdat. Leningrad. Univ., Leningrad, 1961, pp. 3–24. (Russian). RŽ Mat 1962 #4 Б 289.
2. V.M. Babič, *Convergence of series in the ray method of calculating the intensity of wave fronts*, Problems in the Dynamic Theory of the Propagation of Siesmic Waves, no. 5, Izdat. Leningrad. Univ., Leningrad, 1961, pp. 25–35. (Russian) RŽ Mat 1962 #4 Б 290.
3. ———, *Analytic nature of the field of a nonstationary wave in the neighborhood of a caustic*, Problems in the Dynamic Theory of the Propagation of Seismic Waves, no. 5, Izdat. Leningrad. Univ., Leningrad, 1961, pp. 115–144. (Russian) RŽ Mat 1962 #4 Б 291.
4. E.A. Coddington and N. Levinson, *Theory of ordinary differential equations*, McGraw-Hill, New York, 1955; Russian transl., IL, Moscow, 1958. MR **16,** 1022.

Translated by:
M. Lopez-Morillas

ON THE FIELD OF A POINT SOURCE IN THE NEIGHBORHOOD OF AN OPAQUE CIRCULAR CONE. I

B. G. NIKOLAEV

UDC 517 944/947

Abstract. A study is made of the diffraction of the wave field of a point source by an opaque circular cone. The method of finite integral transforms is used to construct a solution of the diffraction problem in a convenient integral form.

Figures: 1. Bibliography: 26 items.

An extensive literature [1–23] has been devoted to the problem of the diffraction of waves described by a single wave equation in the case of conical regions. However, despite the relative abundance of investigations in this field, many questions concerning the description of the physical processes in three-dimensional diffraction problems remain unanswered. It is the aim of the present paper to provide the answer to one of these questions.

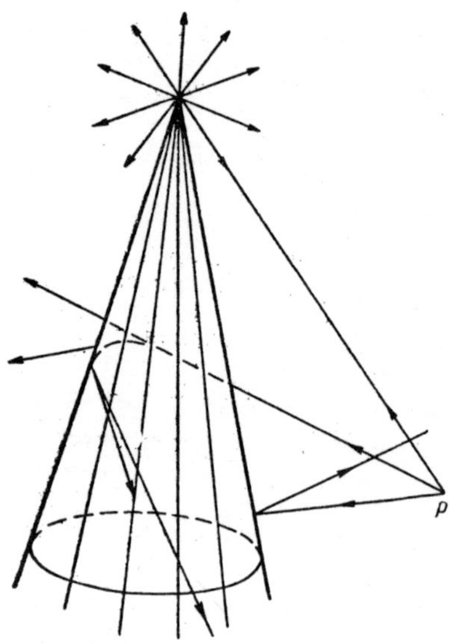

We shall consider a three-dimensional model problem of diffraction from an opaque semi-infinite circular cone. Rays that emanate from a source P reach the

AMS 1970 *subject classifications*. Primary 35J05, 78A45; Secondary 44A15.

surface of the cone (see the figure) and give rise to the rays of the reflected waves and also (in the case of tangential incidence) to rays of waves that propagate along geodesic lines on the cone. Each point of such a line is the starting point of a ray that moves off along the tangent to the geodesic line into the surrounding space. The set of such rays forms the field of the diffracted waves. In addition, diffracted waves are also formed as a result of the scattering of waves from the source by the apex of the cone.

The investigation consists of two parts. In the first part the method of finite integral transforms is used to construct the exact solution of the problem and also establish its validity.

The main aim of the investigation is to distinguish the wave fields of the above three types in the total wave field. The wave fields of the incident and reflected waves can be found (in the approximation of geometrical optics) by the method of steepest descent. The wave scattered from the apex of the cone can be described by a special integral. The diffraction waves from the sides of the cone are separated by means of an identical transformation of the integrand in the expression for the field by the application to the latter of the residue theorem. We shall treat all these questions in the second part of the investigation, which will be prepared shortly.

§1. On the use of finite integral transforms for the solution of boundary value problems of mathematical physics

The traditional approach to problems of diffraction of waves in media with special interfaces that admit separation of the variables in spherical, cylindrical, spheroidal, and various other coordinate systems is to seek the solution in the form of Fourier series in trigonometric functions of an angular coordinate φ. There is usually no difficulty in justifying the process used to construct such solutions. However, difficulties generally arise in practical numerical calculations because of the need to take into account a large number of terms in order to obtain the necessary accuracy at high frequencies and at great distances from the scattering object.

The way out of this difficulty was found by Watson [24], who suggested the replacement of the Fourier series by equivalent contour integrals, the latter being calculated by the contour integration technique.

However, this procedure is not completely consistent; for it would surely be more logical to obtain the solution directly in the form of a contour integral rather than first constructing the solution in the form of a series and then transforming the latter to an integral form. Such a possibility exists and can be realized with comparative ease.

We should point out that this idea is not particularly novel. Indeed, in [25] Tranter mentions finite integral transforms, the analog of Fourier series, and points out that they are suitable for solving problems in the theory of heat conduction. In the fundamental investigation [26], Fourier series are obtained only in the final stage by means of the residue theorem from integral formulas. This merely emphasizes once more that finite integral transforms have remained neglected solely because the value of such transforms has been underestimated.

Accordingly, we shall first consider the expansion of an arbitrary function in a finite integral transform with respect to the solutions of the Sturm-Liouville problem for a second-order linear differential equation of arbitrary form, the aim being to obtain the necessary integral analogs of the Fourier series which will be used to express the solution of the problem of the field of a point source in the neighborhood of an opaque circular cone.

Thus in the interval $a < x < b$ we shall seek an integral formula for the expansion of an arbitrary function $f(x)$ with respect to the solutions of the Sturm-Liouville problem.

In $a < x < b$ we wish to find a function $u(x, \tau)$ that satisfies the equation

$$L[u(x, \tau)] = \tau^2 r(x) u(x, \tau) \tag{1.1}$$

and the boundary conditions

$$\begin{aligned} u(a, \tau)\cos\alpha + u'(a, \tau)\sin\alpha = 0, \\ u(b, \tau)\cos\beta + u'(b, \tau)\sin\beta = 0. \end{aligned} \tag{1.2}$$

Here α and β are fixed parameters, τ takes imaginary values, and

$$L[u] \equiv \frac{d}{dx} p(x) \frac{du}{dx} - q(x) u, \quad q(x) \geqslant 0. \tag{1.3}$$

In this expression $p(x)$, $q(x)$ and $r(x)$ are continuous for $a < x < b$ and do not vanish on this interval, and the function $p(x)$ is continuous together with its first derivative with respect to x.

To solve this problem it is convenient to introduce an auxiliary Cauchy problem for the equation

$$L[v(x, t)] = r(x) \frac{\partial^2 v(x, t)}{\partial t^2} \tag{1.4}$$

with the initial data

$$\begin{aligned} v(x, t)\big|_{t=0} = f(x), \\ \left.\frac{\partial v(x, t)}{\partial t}\right|_{t=0} = 0, \end{aligned} \tag{1.5}$$

where $f(x)$ is an arbitrary continuous function defined in (a, b).

We shall seek the solution of the Cauchy problem in the form of a Mellin integral:

$$v(x, t) = \frac{1}{2\pi i} \int_{\lambda - i\infty}^{\lambda + i\infty} e^{\tau t} w(x, \tau)\, d\tau, \quad \lambda \geqslant 0, \tag{1.6}$$

where

$$w(x, \tau) = \int_0^\infty e^{-\tau t} v(x, t)\, dt. \tag{1.7}$$

Assuming that this integral and also the integral obtained from it by twice differentiating with respect to x converge uniformly for $x \in (a, b)$, and taking into account (1.4), we obtain

$$L[w(x, \tau)] = r(x) \int_0^\infty e^{-\tau t} \frac{\partial^2 v(x, t)}{\partial t^2}\, dt. \tag{1.8}$$

If we now use the condition (1.5) and integrate in (1.8) by parts, we find

$$L[w(x, \tau)] - \tau^2 r(x) w(x, \tau) = -\tau r(x) f(x). \tag{1.9}$$

Augmenting (1.9) with the boundary conditions

$$\begin{aligned} w(a, \tau) \cos\alpha + w'(a, \tau) \sin\alpha &= 0, \\ w(b, \tau) \cos\beta + w'(b, \tau) \sin\beta &= 0, \end{aligned} \tag{1.10}$$

we obtain a closed system of equations for the function $w(x, \tau)$.

For suppose w_1 and w_2 are solutions of the homogeneous equation (1.9) that satisfy the conditions

$$\begin{aligned} w_1(a, \tau) &= \sin\alpha, & w_1'(a, \tau) &= -\cos\alpha, \\ w_2(b, \tau) &= \sin\beta, & w_2'(b, \tau) &= -\cos\beta. \end{aligned} \tag{1.11}$$

Then the Wronskian $W[w_1, w_2]$ of these solutions is independent of x and is a function of τ alone (this can readily be seen, for example, by forming the expression $dW[w_1, w_2]/dx$ for the Wronskian). Let us denote this function by $\omega(\tau)$. It can be shown that $\omega(\tau)$ is an entire function of τ and that the solution of the problem (1.9)–(1.10) can be represented in the form

$$w(x, \tau) = \frac{\tau w_2(x, \tau)}{\omega(\tau)} \int_a^x w_1(y, \tau) r(y) f(y)\, dy$$

$$+ \frac{\tau w_1(x, \tau)}{\omega(\tau)} \int_x^b w_2(y, \tau) r(y) f(y)\, dy. \tag{1.12}$$

Substituting this equation into formula (1.6) (in which we first set $t = 0$) and taking into account the first of the conditions (1.5), we obtain the desired expansion formulas:

$$f(x)=\frac{1}{2\pi i}\int_{\lambda-i\infty}^{\lambda+i\infty} w(x,\tau)\,d\tau,\quad \lambda\geqslant 0, \tag{1.13}$$

$$w(x,\tau)=\frac{\tau w_2(x,\tau)}{\omega(\tau)}\int_a^x w_1(y,\tau)\,r(y)\,f(y)\,dy+$$

$$+\frac{\tau w_1(x,\tau)}{\omega(\tau)}\int_x^b w_2(y,\tau)\,r(y)\,f(y)\,dy. \tag{1.14}$$

The final result can be formulated in the form of the following theorem.

THEOREM 1. *Suppose a function $f(x)$ is continuous and has bounded variation in (a,b). Then the expression*

$$w(x,\tau)=\frac{\tau}{\omega(\tau)}\left[\int_a^x w_1(y,\tau)\,w_2(x,\tau)\,r(y)\,f(y)\,dy\right.$$

$$\left.+\int_x^b w_1(x,\tau)\,w_2(y,\tau)\,r(y)\,f(y)\,dy\right], \tag{1.15}$$

$$\omega(\tau)\equiv W[w_1,w_2],$$

where w_1 and w_2 are the solutions of the problems

$$\begin{gathered}L[w_\nu(x,\tau)]=\tau^2 r(x)\,w_\nu(x,\tau),\quad (\nu=1,2).\\ w_1(a,\tau)=\sin\alpha,\quad w_1'(a,\tau)=-\cos\alpha,\\ w_2(b,\tau)=\sin\beta,\quad w_2'(b,\tau)=-\cos\beta,\\ L[u]\equiv\frac{d}{dx}p(x)\frac{du}{dx}-q(x)\,u,\quad q(x)\geqslant 0\end{gathered} \tag{1.16}$$

defines a regular function $w(x,\tau)$ of τ in the halfplane $\operatorname{Re}\tau=\lambda\geqslant 0$, *and this function satisfies the following boundary conditions with respect to the variable x:*

$$\begin{gathered}w(a,\tau)\cos\alpha+w'(a,\tau)\sin\alpha=0,\\ w(b,\tau)\cos\beta+w'(b,\tau)\sin\alpha=0.\end{gathered} \tag{1.17}$$

In addition, we have the following inversion formula for this integral:

$$f(x)=\frac{1}{2\pi i}\int_{\lambda-i\infty}^{\lambda+i\infty} w(x,\tau)\,d\tau, \tag{1.18}$$

where the integral is taken along any straight line $\mathrm{Re}\,\tau = \lambda \geqslant 0$ parallel to the imaginary axis.

If there are discontinuity points of the first kind, the left side of (1.18) should be replaced by $[f(x+0)+f(x-0)]/2$.

It should be noted that the functions $p(x)$, $q(x)$ and $r(x)$ are continuous in (a,b) and do not vanish in this interval, and $p(x)$ is continuous together with its first derivative with respect to x.

As a first example we give an integral theorem for the expansion of an even, continuous, 2π-periodic function $f(x)$:

$$f(x)=\frac{1}{4\pi i}\int_{\xi-i\infty}^{\xi+i\infty}\frac{e^{\zeta(x+\pi)}}{\sinh\zeta\pi}\,d\zeta\int_0^{2\pi}\cosh\zeta y\,f(y)\,dy,\quad \xi\geqslant 0. \tag{1.19}$$

This formula can be obtained either by using Theorem 1 and taking $p(x)=1$, $q(x)=0$, $a=0$ and $b=2\pi$, or by using the inversion formula for a one-sided Laplace transform and the given properties of $f(x)$.

As a second example we give the integral formula for the expansion of an arbitrary function $f(x;\zeta)$ with respect to modified Legendre functions of the first kind. We shall define functions $w_1(x;\zeta,\tau)$, $w_2(x;\zeta,\tau)$ and $\omega(\zeta,\tau)$ in Theorem 1 by means of the conditions

$$\begin{gathered}\frac{d}{dx}(1-x^2)\frac{dw_\nu}{dx}-\left(\frac{1}{4}-\frac{\zeta^2}{1-x^2}\right)w_\nu=\tau^2 w_\nu\quad(-1<x_1\leqslant x\leqslant 1,\ \nu=1,\,2),\\ w_1(x_1;\,\zeta,\,\tau)=\sin\alpha,\quad w_1'(x_1;\,\zeta,\,\tau)=-\cos\alpha,\\ w_2(1;\,\zeta,\,\tau)=0,\quad w_2'(1;\,\zeta,\,\tau)=-1,\\ \omega(\zeta,\,\tau)=W[w_1(x_1;\,\zeta,\,\tau),\ w_2(x_1;\,\zeta,\,\tau)].\end{gathered} \tag{1.20}$$

As a result we obtain

$$\begin{gathered}w_1(x;\,\zeta,\,\tau)=\left[W\left[P^{+i\zeta}_{-\frac{1}{2}-i\tau}(x_1),\ P^{+i\zeta}_{-\frac{1}{2}-i\tau}(-x_1)\right]\right]^{-1}\\ \times\begin{vmatrix}\cos\alpha P^{+i\zeta}_{-\frac{1}{2}-i\tau}(-x_1)+\sin\alpha\dfrac{d}{dx_1}P^{+i\zeta}_{-\frac{1}{2}-i\tau}(-x_1), & P^{+i\zeta}_{-\frac{1}{2}-i\tau}(-x)\\ \cos\alpha P^{+i\zeta}_{-\frac{1}{2}-i\tau}(x_1)+\sin\alpha\dfrac{d}{dx_1}P^{+i\zeta}_{-\frac{1}{2}-i\tau}(x_1), & P^{+i\zeta}_{-\frac{1}{2}-i\tau}(x)\end{vmatrix},\\ w_2(x;\,\zeta,\,\tau)=-\frac{P^{+i\zeta}_{-\frac{1}{2}-i\tau}(x)}{P^{+i\zeta}_{-\frac{1}{2}-i\tau}(1)},\end{gathered} \tag{1.21}$$

where

$$W\left[P^{+i\zeta}_{-\frac{1}{2}-i\tau}(x_1),\ P^{+i\zeta}_{-\frac{1}{2}-i\tau}(-x_1)\right]=\frac{2}{(1-x_1^2)\,\Gamma\left(-i\zeta+i\tau+\frac{1}{2}\right)\Gamma\left(-i\zeta-i\tau+\frac{1}{2}\right)},$$

$$\omega(\tau,\ \zeta)=-\frac{\cos\alpha P^{+i\zeta}_{-\frac{1}{2}-i\tau}(x_1)+\sin\alpha\frac{d}{dx_1}P^{+i\zeta}_{-\frac{1}{2}-i\tau}(x_1)}{P^{+i\zeta}_{-\frac{1}{2}-i\tau}(1)}, \tag{1.22}$$

and $P_\nu^{-\mu}(x)$ is a modified Legendre function of the first kind.

On the basis of the formulas we have given and the conditions of Theorem 1 we can formulate the following expansion theorem.

THEOREM 2. *Suppose $f(x;\zeta)$ is continuous in x and has bounded variation in $-1 < x_1 \leqslant x \leqslant 1$ and ζ is a parameter that takes imaginary values. Then the expression*

$$w(x;\ \zeta,\ \tau)=\frac{1-x_1^2}{2}\cdot\frac{\tau\Gamma\left(-i\zeta+i\tau+\frac{1}{2}\right)\Gamma\left(-i\zeta-i\tau+\frac{1}{2}\right)}{\cos\alpha P^{+i\zeta}_{-\frac{1}{2}-i\tau}(x_1)+\sin\alpha\frac{d}{dx_1}P^{+i\zeta}_{-\frac{1}{2}-i\tau}(x_1)}$$

$$\times\left[\int_{x_1}^{x}Q(y;\ \zeta,\ \tau)\,P^{+i\zeta}_{-\frac{1}{2}-i\tau}(x)\,f(y;\ \zeta)\,dy\right.$$

$$\left.+\int_{x}^{1}Q(x;\ \zeta,\ \tau)\,P^{+i\zeta}_{-\frac{1}{2}-i\tau}(y)\,f(y;\ \zeta)\,dy\right], \tag{1.23}$$

where

$$Q(x;\ \zeta,\ \tau)=\begin{vmatrix}\cos\alpha P^{+i\zeta}_{-\frac{1}{2}-i\tau}(-x_1)+\sin\alpha\frac{d}{dx_1}P^{+i\zeta}_{-\frac{1}{2}-i\tau}(-x_1), & P^{+i\zeta}_{-\frac{1}{2}-i\tau}(-x),\\ \cos\alpha P^{+i\zeta}_{-\frac{1}{2}-i\tau}(x_1)+\sin\alpha\frac{d}{dx_1}P^{+i\zeta}_{-\frac{1}{2}-i\tau}(x_1), & P^{+i\zeta}_{-\frac{1}{2}-i\tau}(x)\end{vmatrix} \tag{1.24}$$

defines a regular function $w(x;\zeta,\tau)$ of τ in the halfplane $\mathrm{Re}\,\tau=\lambda\geqslant 0$, *this function satisfying the following boundary conditions with respect to x:*

$$\begin{aligned}&w(x_1;\ \zeta,\ \tau)\cos\alpha+w'(x_1;\ \zeta,\ \tau)\sin\alpha=0,\\ &w(1;\ \zeta,\ \tau)=0,\quad w'(1;\ \zeta,\ \tau)=-1.\end{aligned} \tag{1.25}$$

In addition, we have the inversion formula

$$f(x;\zeta)=\frac{1}{2\pi i}\int_{\lambda-i\infty}^{\lambda+i\infty} w(x;\zeta,\tau)\,d\tau, \tag{1.26}$$

where the integral is along any straight line $\mathrm{Re}\tau = \lambda \geqslant 0$ parallel to the imaginary axis. In the case of discontinuity points of the first kind, the left side in the last equation must be replaced by $[f(x+0;\zeta)+f(x-0;\zeta)]/2$.

Formula (1.19) and also the results of Theorem 2 can be used to construct the exact solution of the model problem of the field of a point source in the neighborhood of an opaque circular cone.

§2. Statement of the problem of diffraction of waves from an opaque circular cone and the method of solution of this problem

The mathematical formulation of the problem of diffraction from an opaque circular cone is as follows.

In a coordinate system (r,θ,φ) there is given a region D $(r \geqslant 0, 0 \leqslant \theta < \theta_1, 0 < \varphi \leqslant 2\pi)$ filled with a medium characterized by a value of the wave number k and a region that is the complement of D^* in the complete space. This second region is an opaque semi-infinite circular cone with opening angle $2\pi - 2\theta_1$. At the point $P(r_0,\theta_0,0)$ of D there is a source of monochromatic spherical waves that oscillate in time in accordance with the law $\exp(-i\omega t)$.

The wave processes in the medium satisfy the inhomogeneous Helmholtz equation

$$\Delta u(r,\theta,\varphi)+k^2u(r,\theta,\varphi)=-\frac{\delta(r-r_0)\,\delta(\theta-\theta_0)\,\delta(\varphi)}{r^2\sin\theta}, \tag{2.1}$$

where $\delta(x)$ is the delta function provided that a boundary condition of mixed kind is satisfied at the boundary $\theta=\theta_1$:

$$\left[\cos\alpha u(r,\theta,\varphi)+\sin\alpha\frac{du(r,\theta,\varphi)}{d(\cos\theta)}\right]_{\theta=\theta_1}=0, \tag{2.2}$$

which includes both the Dirichlet boundary condition ($\alpha=0$) and the Neumann boundary condition ($\alpha=\pi/2$).

In addition, it is assumed that the desired function $u(r,\theta,\varphi)$ is even in the variable φ:

$$u(r,\theta,-\varphi)=u(r,\theta,\varphi) \tag{2.3}$$

and is also periodic in φ:

$$u(r,\theta,\varphi+2\pi)=u(r,\theta,\varphi). \tag{2.4}$$

Under these assumptions we seek a wave field $U = u(r, \theta, \varphi)\exp(-i\omega t)$ that satisfies the radiation condition[1] for $r > r_0$ and is finite on the axis $\theta = 0$ right up to the apex $r = 0$ of the cone.

In accordance with formula (1.19) we shall represent the solution of the problem (2.1)–(2.4) in the form

$$u(r, \theta, \varphi) = \frac{1}{4\pi i} \int_{\xi - i\infty}^{\xi + i\infty} \frac{e^{\zeta(\varphi+\pi)}}{\sinh \zeta\pi} v(r, \theta; \zeta)\, d\zeta, \tag{2.5}$$

where

$$v(r, \theta; \zeta) = \int_0^{2\pi} \cosh \zeta y\, u(r, \theta, y)\, dy. \tag{2.6}$$

Integrating (2.1) with respect to φ over $(0, 2\pi)$ with the weight $\cosh \zeta\varphi$ and using the condition (2.4), we obtain the following equation from (2.6) for the function $v(r, \theta; \zeta)$:

$$\frac{1}{r^2}\frac{\partial}{\partial r} r^2 \frac{\partial v}{\partial r} + k^2 v + \frac{1}{r^2 \sin\theta}\frac{\partial}{\partial \theta} \sin\theta \frac{\partial v}{\partial \theta} + \frac{\zeta^2}{r^2 \sin^2\theta} v = -\frac{\delta(r - r_0)\,\delta(\theta - \theta_0)}{r^2 \sin\theta} + h(r, \theta; \zeta), \tag{2.7}$$

where

$$h(r, \theta; \zeta) = \frac{\zeta \sinh 2\pi\zeta\, u\,|_{\varphi=0} - (\cosh 2\pi\zeta - 1)\left.\frac{\partial u}{\partial \varphi}\right|_{\varphi=0}}{r^2 \sin\theta}. \tag{2.8}$$

In addition, it follows from the boundary condition (2.2) that $v(r, \theta; \zeta)$ satisfies a similar condition for $\theta = \theta_1$:

$$\left[\cos\alpha\, v(r, \theta; \zeta) + \sin\alpha \frac{\partial v(r, \theta; \zeta)}{\partial(\cos\theta)}\right]_{\theta=\theta_1} = 0. \tag{2.9}$$

Finally, the uniqueness theorem for the solution of the problem (2.1)–(2.4) imposes restrictions on $v(r, \theta; \zeta)$ due to the radiation condition and the condition of finiteness at $\theta = 0$ and $r = 0$.

Since the problem (2.7)–(2.9) is linear, the function $v(r, \theta; \zeta)$ can be represented in the form $v(r, \theta; \zeta) = v_1(r, \theta; \zeta) + v_2(r, \theta; \zeta)$, where $v_1(r, \theta; \zeta)$ is the disturbance due to the first term on the right side of (2.7) and $v_2(r, \theta; \zeta)$ depends only on the value of the second term. It is readily seen that $v_2(r, \theta; \zeta)$ makes no contribution to the field $u(r, \theta, \varphi)$ defined by (2.5); for since $h(r, \theta; \zeta)$ is proportional to $\sinh \pi\zeta$ it follows that

[1] In accordance with this principle the total wave field of the problem of diffraction from a cone must not contain disturbances in the form of waves coming in from infinity.

the disturbance $v_2(r,\theta;\zeta)$ corresponding to this function is also proportional to $\sinh \pi\zeta$, and hence the integrand in the integral (2.5) of this disturbance will not contain singularities and the integral (2.5) of $v_2(r,\theta;\zeta)$ will vanish by Cauchy's theorem. Therefore when we speak of the function $v(r,\theta;\zeta)$ we shall henceforth assume that it is determined from the solution of the problem

$$\frac{1}{r^2}\frac{\partial}{\partial r}r^2\frac{\partial v}{\partial r}+k^2v+\frac{1}{r^2\sin\theta}\frac{\partial}{\partial\theta}\sin\theta\frac{\partial v}{\partial\theta}+\frac{\zeta^2}{r^2\sin^2\theta}v = -\frac{\delta(r-r_0)\,\delta(\theta-\theta_0)}{r^2\sin\theta}, \tag{2.10}$$

$$\left[\cos\alpha\, v(r,\theta;\zeta)+\sin\alpha\frac{\partial v(r,\theta;\zeta)}{\partial(\cos\theta)}\right]_{\theta=\theta_1}=0, \tag{2.11}$$

where $v(r,\theta;\zeta)$ satisfies the radiation condition for $r > r_0$ and the condition of finiteness at $\theta = 0$ and $r = 0$.

In accordance with the expansion theorem (see §1, Theorem 2) for the modified Legendre functions of the first kind the solution of the problem (2.10)–(2.11) can be represented in the form

$$v=\frac{1}{4\pi i}\int\limits_{\lambda-i\infty}^{\lambda+i\infty}\frac{(1-x_1^2)\Gamma\left(-i\zeta+i\tau+\frac{1}{2}\right)\Gamma\left(-i\zeta-i\tau+\frac{1}{2}\right)w(r,x;\zeta,\tau)}{\cos\alpha P^{+i\zeta}_{-\frac{1}{2}-i\tau}(x_1)+\sin\alpha\frac{d}{dx_1}P^{+i\zeta}_{-\frac{1}{2}-i\tau}(x_1)}\tau\,d\tau, \tag{2.12}$$

$$w(r,x;\zeta,\tau)=P^{+i\zeta}_{-\frac{1}{2}-i\tau}(x)\int\limits_{x_1}^{x}Q(y;\zeta,\tau)\,v(r,y;\zeta)\,dy + Q(x;\zeta,\tau)\int\limits_{x}^{1}P^{+i\zeta}_{-\frac{1}{2}-i\tau}(y)\,v(r,y;\zeta)\,dy, \tag{2.13}$$

where $x = \cos\theta$, $x_1 = \cos\theta_1$ and

$$Q(y;\zeta,\tau) = \begin{vmatrix} \cos\alpha P^{+i\zeta}_{-\frac{1}{2}-i\tau}(-x_1)+\sin\alpha\frac{d}{dx_1}P^{+i\zeta}_{-\frac{1}{2}-i\tau}(-x_1), & P^{+i\zeta}_{-\frac{1}{2}-i\tau}(-x) \\ \cos\alpha P^{+i\zeta}_{-\frac{1}{2}-i\tau}(x_1)+\sin\alpha\frac{d}{dx_1}P^{+i\zeta}_{-\frac{1}{2}-i\tau}(x_1), & P^{+i\zeta}_{-\frac{1}{2}-i\tau}(x) \end{vmatrix} \tag{2.14}$$

To find which condition must be satisfied by the function $w(r,x;\zeta,\tau)$ in (2.13), we integrate (2.10) with respect to $x = \cos\theta$ in accordance with (2.13). The outcome is

$$\frac{1}{r^2}\frac{d}{dr}r^2\frac{dw}{dr}+\left[k^2+\frac{1}{r^2}\left(\tau^2+\frac{1}{4}\right)\right]w=-\frac{\delta(r-r_0)}{r^2} \times\begin{cases} P^{+i\zeta}_{-\frac{1}{2}-i\tau}(x)\,Q(x_0;\zeta,\tau), & (x_1<x_0<x) \\ Q(x;\zeta,\tau)\,P^{+i\zeta}_{-\frac{1}{2}-i\tau}(x_0), & (x<x_0<1) \end{cases} + g(r,x;\zeta,\tau), \tag{2.15}$$

where

$$g(r,\ x;\ \zeta,\ \tau)\equiv\frac{1}{r^2}\left\{P^{+i\zeta}_{-\frac{1}{2}-i\tau}(x)\left[(1-x^2)\left(Q\frac{\partial v}{\partial x}-v\frac{\partial Q}{\partial x}\right)\right]_{x=x_1}^{x=x}\right.$$

$$\left.+Q(x;\ \zeta,\ \tau)\left[(1-x^2)\left(P^{+i\zeta}_{-\frac{1}{2}-i\tau}(x)\frac{\partial v}{\partial x}-v\frac{\partial}{\partial x}P^{+i\zeta}_{-\frac{1}{2}-i\tau}(x)\right)\right]_{x=x}^{x=1}\right\}. \tag{2.16}$$

Allowance for the boundary conditions (2.11) and also the easily verified identities

$$\left[\cos\alpha\, Q+\sin\alpha\frac{\partial Q}{\partial x}\right]_{x=x_1}=0,$$

$$W\left[P^{+i\zeta}_{-\frac{1}{2}-i\tau}(x_1),\ P^{+i\zeta}_{-\frac{1}{2}-i\tau}(-x_1)\right]$$

$$=\frac{2}{(1-x_1^2)\Gamma\left(-i\zeta+i\tau+\frac{1}{2}\right)\Gamma\left(-i\zeta-i\tau+\frac{1}{2}\right)} \tag{2.17}$$

enables us to transform the function $g(r,x;\zeta,\tau)$ in (2.16) to the form

$$g(r,\ x;\ \zeta,\ \tau)=-\frac{2}{r^2}\left(\cos\alpha P^{+i\zeta}_{-\frac{1}{2}-i\tau}(x_1)+\sin\alpha\frac{\partial}{\partial x_1}P^{+i\zeta}_{-\frac{1}{2}-i\tau}(x_1)\right)$$

$$\times\frac{v(r,\ x;\ \zeta,\ \tau)}{\Gamma\left(-i\zeta+i\tau+\frac{1}{2}\right)\Gamma\left(-i\zeta-i\tau+\frac{1}{2}\right)}. \tag{2.18}$$

The term $g(r,x;\zeta,\tau)$ in (2.15) corresponds to the part of the function w which makes a vanishing contribution to the expression for v determined by (2.12); that is, this term contains as a factor a quantity that eliminates the singularity in the integrand of (2.12). This observation enables us to use a simpler equation to find the function $w(r,x;\zeta,\tau)$:

$$\frac{1}{r^2}\frac{d}{dr}r^2\frac{dw}{dr}+\left[k^2+\left(\tau^2+\frac{1}{4}\right)\frac{1}{r^2}\right]w$$

$$=-\frac{\delta(r-r_0)}{r^2}\begin{cases}P^{+i\zeta}_{-\frac{1}{2}-i\tau}(x)\,Q(x_0;\ \zeta,\ \tau), & (x_1<x_0<x)\\ Q(x;\ \zeta,\ \tau)\,P^{+i\zeta}_{-\frac{1}{2}-i\tau}(x_0), & (x<x_0<1)\end{cases} \tag{2.19}$$

which is compatible with the radiation condition for $r>r_0$ and the condition of finiteness $r=0$.

The solution fo the problem associated with (2.19) has the form

$$w = \frac{\pi i}{2}\left[\frac{J_{-i\tau}(kr)}{\sqrt{r}}\int_0^r \frac{H^{(1)}_{-i\tau}(k\rho)\, q(\rho, x; \zeta, \tau)\, \rho^2 d\rho}{\sqrt{\rho}} + \frac{H^{(1)}_{-i\tau}(kr)}{\sqrt{r}}\int_r^\infty \frac{J_{-i\tau}(k\rho)\, q(\rho, x; \zeta, \tau)\, \rho^2 d\rho}{\sqrt{\rho}}\right], \tag{2.20}$$

where

$$q(r, x; \zeta, \tau) = -\frac{\delta(r - r_0)}{r^2}\begin{cases} P^{+i\zeta}_{-\frac{1}{2}-i\tau}(x)\, Q(x_0; \zeta, \tau), & (x_1 < x_0 < x), \\ Q(x; \zeta, \tau)\, P^{+i\zeta}_{-\frac{1}{2}-i\tau}(x_0), & (x < x_0 < 1). \end{cases} \tag{2.21}$$

After the integration has been performed in (2.20) and (2.21) we obtain the following expression for $w(r, x; \zeta, \tau)$:

$$w(r, x; \zeta, \tau) = w_1(r, r_0; \tau) \cdot w_2(x, x_0; \zeta, \tau), \tag{2.22}$$

where

$$w_1(r, r_0; \tau) = \frac{\pi i}{2}\begin{cases} \frac{J_{-i\tau}(kr)}{\sqrt{r}}\,\frac{H^{(1)}_{-i\tau}(kr_0)}{\sqrt{r_0}}, & (r < r_0), \\ \frac{H^{(1)}_{-i\tau}(kr)}{\sqrt{r}}\,\frac{J_{-i\tau}(kr_0)}{\sqrt{r_0}}, & (r > r_0); \end{cases} \tag{2.23}$$

$$w_2(x, x_0; \zeta, \tau) = \begin{cases} P^{+i\zeta}_{-\frac{1}{2}-i\tau}(x)\, Q(x_0; \zeta, \tau), & (x_1 < x_0 < x), \\ Q(x; \zeta, \tau)\, P^{+i\zeta}_{-\frac{1}{2}-i\tau}(x_0), & (x < x_0 < 1); \end{cases} \tag{2.24}$$

$$Q(y; \zeta, \tau) = \begin{vmatrix} \cos\alpha P^{+i\zeta}_{-\frac{1}{2}-i\tau}(-x_1) + \sin\alpha \frac{d}{dx_1} P^{+i\zeta}_{-\frac{1}{2}-i\tau}(-x_1), & P^{+i\zeta}_{-\frac{1}{2}-i\tau}(-y) \\ \cos\alpha P^{+i\zeta}_{-\frac{1}{2}-i\tau}(x_1) + \sin\alpha \frac{d}{dx_1} P^{+i\zeta}_{-\frac{1}{2}-i\tau}(x_1), & P^{+i\zeta}_{-\frac{1}{2}-i\tau}(y) \end{vmatrix}. \tag{2.25}$$

Substituting the values found for the functions v and w into the formula for $U = u(r, \theta, \varphi)\exp(-i\omega t)$, we obtain the final form of the solution of the problem of diffraction of the field of a point source in the neighborhood of an opaque cone:

$$U=\frac{e^{-i\omega t}}{(4\pi i)^2}\int_{\xi-i\infty}^{\xi+i\infty}\frac{e^{\zeta(\varphi+\pi)}\cdot}{\sinh\zeta\pi}d\zeta$$

$$\times\int_{\lambda-i\infty}^{\lambda+i\infty}\frac{(1-x_1^2)\Gamma\left(-i\zeta+i\tau+\frac{1}{2}\right)\Gamma\left(-i\zeta-i\tau+\frac{1}{2}\right)}{\cos a P^{+i\zeta}_{-\frac{1}{2}-i\tau}(x_1)+\sin a\frac{d}{dx_1}P^{+i\zeta}_{-\frac{1}{2}-i\tau}(x_1)}$$

$$\times w_1(r,\ r_0;\ \tau)\, w_2(x,\ x_0;\ \zeta,\ \tau)\,\tau d\tau, \qquad (2.26)$$

where w_1 and w_2 are determined by (2.23) and (2.24).

To justify this solution it is sufficient to transform it by means of the Cauchy residue theorem to the form of double series and to use the estimates for the cylindrical and modified Legendre functions that follow from the asymptotic representations of these functions. These estimates show that the series can be differentiated an arbitrary number of times under the summation sign for $r \neq r_0$ and $\theta \neq \theta_0$ and, in particular, that they satisfy the Helmholtz equation.

In addition, the form of the expansions

$$H^{(1)}_{p-\frac{1}{2}}(kr)=\sqrt{\frac{2}{\pi kr}}\,e^{i\left(kr-\frac{p\pi}{2}\right)}\left[1+0\left(\frac{1}{kr}\right)\right],$$

$$J_{p-\frac{1}{2}}(kr)=\frac{\left(\frac{kr}{2}\right)^{p-1/2}}{\Gamma\left(p+\frac{1}{2}\right)}\left[1+0\left(\frac{k^2r^2}{4}\right)\right] \qquad (2.27)$$

for the cylindrical functions $H_\nu^{(1)}(z)$ and $J_\nu(z)$ shows that the function $U(r,\theta,\varphi,t)$ satisfies the radiation condition for $r > r_0$ and the condition of finiteness at $\theta = 0$ and $r = 0$.

Finally, differentiating with respect to θ, we can readily show that the boundary condition (2.2) is satisfied at the surface $\theta = \theta_1$ of the cone.

BIBLIOGRAPHY

1. H. Carslaw, *The scattering of sound waves by a cone*, Math. Ann. **75** (1914), 133–147; errata, 592.
2. H. Buchholz, *Einige Gesetzmassigkeiten über die Bewegung elektromagnetischer Wellen in einem Kegelhorn*, Arch. Electrotechnik **40** (1950/52), 346–362.
3. ———, *Elektromagnetische Felder mit axialer Symmetrie bei Begrenzungen durch Kegel und Kugel*, Arch. Electrotechnik **45** (1960), 27–48.
4. L.B. Felsen, *Backscattering from wide-angle and narrow-angle cones*, J. Appl. Phys. **26** (1955), 138–151. MR **16**, 773.
5. ———, *Alternative field representations in regions bounded by spheres, cones, and planes*, IRE Trans. Antennas and Propagation **AP-5** (1957), 109–121.
6. ———, *Plane-wave scattering by small-angle cones*, IRE Trans. Antennas and Propagation **AP-5** (1957), 121–129.
7. ———, *Asymptotic expansion of the diffracted wave for a semi-infinite cone*, IRE Trans. Antennas and Propagation **AP-5** (1957), 402–404.

8. ———, *Radiation from ring sources in the presence of a semi-infinite cone*, IRE Trans. Antennas and Propagation **AP-7** (1959), 168–180.
9. S.N. Belonosov, Trudy Rjazansk. Radiotehn. Inst. **1** (1956), 203–205. (Russian)
10. K.M. Siegel, J.W. Crispin and C.E. Schensted, *Electromagnetic and acoustical scattering from a semi-infinite cone*, J. Appl. Phys. **26** (1955), 309–313. MR **16,** 1181.
11. L.L. Bailin and S. Silver, *Exterior electromagnetic boundary value problems for spheres and cones*, IRE Trans. Antennas and Propagation **AP-4** (1956), 5–16.
12. G. Held and G. Hasserdjian, *Surface fields produced by a slot on a cone*, IRE Trans. Antennas and Propagation AP-5 (1957), 398–399.
13. J.B. Keller, *Backscattering from a finite cone*, IRE Trans. Antennas and Propagation **AP-8** (1960), 175–182.
14. ———, *Backscattering from a finite cone—comparison of theory and experiment*, IRE Trans. Antennas and Propagation AP-9 (1961), 411–412.
15. A.S. Gorjainov, *Diffraction of a plane electromagnetic wave propagating along the axis of a cone*, Radiotehnika i Èlektr. **6** (1961), 47–57 = Radio Engrg. Electron. Phys. **6** (1961), 39–48.
16. L. Kraus and L.M. Levine, *Diffraction by an elliptic cone*, Comm. Pure Appl. Math. **14** (1961), 49–68. MR **22** #10554.
17. P.V. Coĭ, *Diffraction of (short) sound waves at an obstacle (cylinder, sphere, cone and plane)*, Prikl. Mat. Meh. **25** (1961), 365–369 = J. Appl. Math. Mech. **25** (1961), 536–544. MR **24** #B2277.
18. A. Kumar and R. Chatterjee, *Electromagnetic waves on a semi-infinite conically tapered circular dielectric rod*, J. Inst. Telecomm. Engrs. (India) **8** (1962), no. 4, 171–178. EEA **66** #7276.
19. E.M. Kennaugh and D.L. Moffatt, *On the axial echo area of the cone-sphere shape*, Proc. IRE **50** (1962), 199.
20. F.H. Northover, *The diffraction of electromagnetic waves around a finite, perfectly conducting cone*, Quart. J. Mech. Appl. Math. **15** (1962), 1–9. MR **25** #921.
21. V.A. Borovikov, *Reduction of certain three-dimensional diffraction problems to the Dirichlet problem for the Laplace equation*, Dokl. Akad. Nauk SSSR **144** (1962), 527–530 = Soviet Physics Dokl. **7** (1962), 385–387.
22. G.A. Grinberg, *Solution of the static and wave problems for screens shaped like a cone or part of a conical surface*, Ž. Tehn. Fiz. **33** (1963), 1392–1394 = Soviet Physics Tech. Phys. **8** (1964), 1035–1036. MR **32** #8679.
23. V.B. Poručikov, *Solution of the problem of the diffraction of an acoustic wave by a cone*, Prikl. Mat. Meh. **32** (1968), 319–323 = J. Appl. Math. Mech. **32** (1968), 312–316.
24. G. Watson, *The diffraction of electric waves by the earth*, Proc. Roy. Soc. London Ser. A **95** (1918), 83–99.
25. C.J. Tranter, *Integral transforms in mathematical physics*, Methuen, London, 1951; Russian transl., GITTL, Moscow, 1956, 12–13. MR 13, 127.
26. E.C. Titchmarsh, *Eigenfunction expansions associated with second-order differential equations*. Part 1, Clarendon Press, Oxford, 1946; Russian transl., IL, Moscow, 1960. MR **8,** 458; **22** #9662.

Translated by:
J. R. Barbour